SIMON V

KINGDOM OF
VIKINGS

THE RISE AND
FALL OF NORWAY

ISBN 978-1-5272-8017-5 (Paperback)
ISBN 978-1-5272-8018-2 (E-book)

Cover art by PaperTrue Ltd.

Printed in United Kingdom

Visit www.strategoshistory.com

Cover image: Harald Hardrada window at Kirkwall Cathedral by Colin Smith, under the Creative Commons Attribution-Share Alike 2.0 Generic License. Background removed and colour retouched.

*To Svein-Ivar Kr. Herland,
thank you for all our conversations,
my friend.*

"Cattle die, kinsmen die, all men are mortal. But words of a legacy never perish, nor a noble name."

– Norse Proverb

"We know that the Norwegians are rough and warlike, and it is dangerous to carry hostility to their doors."

– Anund Jacob of Sweden

Contents

Introduction

In 1230 AD, on the obscure island of Iceland, a middle aged Norse huscarl sat down and began to write. His name was Snorri Sturlason. Though he was a warrior, he was also a poet and author. He enjoyed a wealth of knowledge, for he had collected all the oral lore concerning the epic tales of his forefathers. Snorri knew this history better than anyone. Gifted by lyrical flair and good memory, he had now been requested by the King of Norway to write the epic sagas of the Kings. If he could write down the origins of the Vikings, the great deeds and acts they accomplished while still alive and the forging of their Kingdoms – he would immortalise them. These stories are rich in both charm and dread. They follow the valiant Kings of old who dwelled in the Northern mountains and seas, and decided to carve out a Kingdom there, and call it the land of the Northmen. Through the beauty of Norse rhyming, the stories of these men had so far been recorded in an endless amount of poems. Snorri's innovation was to translate them into written sagas. This book's innovation is to translate the sagas into a modern recollection of how the Vikings and their descendants forged the throne of ancient Norway.

The story of how the Kingdom of Norway was forged is perhaps one of the most fascinating, adventurous, and riveting stories in history. It weaves together the lives of twelve grand figures; spans five centuries of war and peace; and covers the edges of Siberia to the woodlands of Canada, the icebergs of the Arctic to the desert sands of Africa. Norway was not built in a day, nor by one single person. It was forged slowly, being borne out of the wild Viking era, and hammered into maturity through the middle ages.

Norway is an ancient land. Its origins are clouded by myth and mystery. Its tall and majestic mountains have witnessed many untold events. But the mountains do not share their secrets, so it is up to us to uncover them. We must therefore collaborate with both medieval and contemporary historians; Viking poets and current-day archaeologists; to illuminate the truths of how this Northern outpost was made a Kingdom. What we are left with is a comprehensive set of life stories, told in a captivating and visually compelling manner, that does justice to both the medieval sagas and the scholarship of history.

✴ Introduction ✴

This is a real game of thrones. The titans described in this book all left a unique footprint on this Northern, Viking kingdom – but they made major sacrifices to achieve it. The story of the Kingdom of Norway is riddled with cunning geopolitics, fierce battles, notorious raiding, and bitter betrayal. However, it also displays grand statesmanship, unyielding loyalty and friendship, love affairs and powerful marriages, sacrificial courage and stirring valour on the battlefield. Thus, the following story fills us with a range of emotions - dread, charm, and inspiration – while also sparking our intellectual curiosity.

Odin's cradle

The Norwegian archaeologist Thor Heyerdahl was known for his unconventional approach to history and myths. He stirred a lot of interest when he claimed that the story of the Norse god Odin – told in Snorri Sturlason's *Heimskringla* and *Poetic Edda* – was real, not fiction. He therefore took Snorri seriously, the same way Snorri took the old legend of Odin, seriously.

According to Snorri, Odin was not a fantastical, invented figure – but a real person. Odin was the King of an ancient people, residing somewhere in modern-day Caucasus, with the pride of their tribe being the splendid Aas citadel (Åsgard). With the invasion of Roman legions, they fled and migrated North, eventually settling in Scandinavia. Odin was wise and strong, and when he died, he was so dearly loved that his followers began worshipping his spirit. This began the Odin-cult, or the Norse religion (also called *Asatru* – the faith of Asgard). *Norsemen,* therefore, refers to these Scandinavian peoples who followed the Norse religion and culture.

Odin's ruling dynasty would be called the *Ynglings* (modern scholars suspect the name originates from *Scythlings,* or Scythians). All members of Odin's deified family became rulers over the Scandinavian lands, forming different Norse tribes (Jutes, Zealanders, Suedi, Gothic, Trondic, etc). Some went South-West to settle in modern-day Denmark[1]. Some stayed in Uppsala in Sweden[2]. Others decided to cross the awe-inspiring Northern mountains and settle along the stretched, barren coastline of modern-Norway[3]. Here they dwelled for centuries – developing villages, merchant towns and agricultural farmlands.

---

[1] Here settled the old tribes of Jutes, Zealanders, and others. While fighting the Romans, they came under lordship of King Dan, a descendant of Odin. Dan could have given the land its name – *Danmark* (Denmark) – as *mark* indicates border.

[2] Suedi and Gothic (or Geat/Gaut) tribes.

[3] Notably the Tronds, but also the tribes of Maere, the Sogn, the Hords, and others.

✴ Introduction ✴

To the Romans, this Northern region of Scandinavia was both mysterious and obscure. However, they knew that its inhabitants had much in common with other Germanic peoples they had encountered in central Europe. Tacitus, the Roman historian, investigated the Germanic tribes in 98 A.D and wrote down his assessments. He emphasised that they had a strong culture of individual freedom and communal customs. "*...and good customs are here more effectual than good laws elsewhere,*" Tacitus noted. They were deeply suspicious of Kings and absolute rulers. In their rule of thought, the King ought to serve the people, not the other way around. However, should there be war, they regarded it as a duty and an honour to die fighting for their leader.

When the Romans neared the borders of modern-day Denmark, they clashed fiercely with the Jutes, who then rallied the other Norse tribes under the lordship of King Dan (who most likely gave Denmark its name) and assaulted the Romans. The Romans fled and would never return.

However, though the Romans would not return to Scandinavia, the Scandinavians returned to them. In the 5th century A.D, Saxons and Anglers - two tribes hailing, according to the legend, from Odin's people – swept across Britannia and claimed it for themselves. They would be known as the Anglo-Saxons and would rule England for hundreds of years. Other Germanic tribes – Franks, Goths, Suedi, Vandals – stormed through the Western Roman Empire and destroyed it. The classical world was shattered. Europe began a new era.

The Franks settled in modern-day France and would inherit a lot of traditions, laws, and institutions from the ancient Romans. In many ways, they regarded themselves of the inheritors of the Roman legacy. By the early 700s, their King, Charles the Great (or *Charlemagne* from the Latinised Carlo Magnus), created a vast Christian Empire that pushed back into the old Northern territories – perhaps *too* deep. As his Frankish troops ventured into Denmark, they made the same mistake the Romans had done seven centuries before and clashed fiercely with the pagan peoples. The Franks called them *Nortmenn* or Northmen – a Nordic Germanic people who had never left their ancestral lands nor changed their customs. Very little was known about them.

So, who were these Norse peoples? The Franks knew they were very shrewd tradesmen. They had used to trade with ancient Romans before, but also with distant peoples (for instance, a Buddha statue dating 500 A.D has been found in modern-day Sweden, suggesting a complex network of trade routes). They also fought fiercely with other Norse clans to expand their holdings, avenge their ancestors, or raid each other's wealth. The only cause that seemed to unite them was when a foreign intruder (non-Norse) invaded – then, they quickly rallied to inflict terrible madness upon their common foe.

✷ Introduction ✷

Ancient Norway

The most well-known area to other Europeans was Denmark. The other territories were obscure and strange. However, they did know of a very convenient naval route along Scandinavia's North Sea coastline. It led all the way up to the Arctic circle, and was sheltered from the rough North Sea by an endless amount of archipelagos and islands. Bold merchants could venture up here to trade animal furs and pelts, and walrus ivory – all quite valuable commodities in Central Europe. The trade route was allegedly known as the *North Way*. The inhabitants of the coastline were reserved, but hard knocked.

The ancient historian Jordanes makes an account of these peoples in the context of an attack on Denmark from the North. He claimed that they were then ruled by one chief – Rodulf – and were exceptionally tall. "[They were] *taller and wilder than the Germani...*[fighting] *with animal ferocity,*" Jordanes tells. Among the tribes united under Rodulf were the Rugi, who are also mentioned as having participated in the invasions of Roman lands and fought under the notorious Attila the Hun.

These Northern peoples lived in harsh environments. Winters were long and dark, summers were short. All that mattered was how many winters a person could survive before death. The terrain was often rugged and desolate, with very little arable land to sustain larger populations. They therefore often set out on the vast North Sea in search of good fishing prospects, but the Sea was no haven. The Nordic climate caused freezing winds and violent storms, and if one drifted too far North, one risked hitting icebergs or getting lost at sea. Navigation and seamanship were not just useful skills – they were a matter of life or death. Above all, one had to take courage, think rationally and never buckle in weakness. They were bred by nature to become tough and terse. It was survival of the fittest.

The North Way coastline was shared among various chiefs and petty kings hailing from the Yngling dynasty (as the sagas relate). The Far North was held by the peoples of Haloys (Halogaland); the mid-section of the coast was held by the mighty Tronds – the most powerful and populous tribe of the region (Trondelag or Trondheim, meaning Trond's home); south to them were the peoples of Maere (current-day Møre); then came the Sogn clan; the Hords (Hordaland); the Rygs/Rugi (Rogaland); the Agds (Agder); and then around the land of Viken lay various petty kingdoms and chiefdoms (Vestfold, Ringerike, Hedemark). Between Viken and Trondelag lay the Uplands, home to various petty kings of old. This was the ancient division of the lands of the North Way – Ancient Norway.

✷ Introduction ✷

Middle Earth – the World of the Viking

"These ships are not loaded with cargo but with hostile wild men...I am not frightened that these pirates will hurt me. But it pains my heart to think that they have sometimes, in my lifetime, dared to attack our coasts, and I am struck with fear to imagine what evil they will bring to my successors and their subjects."

– Charlemagne when witnessing Viking ships entering his harbour.

In 793 A.D, a group of Norse raiders, known as *Vikings*, attacked and plundered the famous monastery of Lindisfarne – one of the finest centres of Christendom in Britannia. For the Christians, it was an unbelievable sacrilege. For the Vikings, an easy loot. Since the monasteries were packed with gold, yet undefended, Vikings repeatedly returned to sack them. This was the beginning of the so-called "Viking Age" – a period that took its name from the Norse word *viking*, meaning raider. Unfortunately for the Europeans, this age would last for a long time.

European kingdoms struck back with force, but instead of intimidating the Vikings, they only provoked them. Returning to their shores in year 865 were not bands of brigands – but entire armies. The so-called *Great Heathen Army*, led chiefly by Ivar the Boneless, invaded England and destroyed anything in its path. For a moment, they held almost all of England, making York their headquarters. Though they were subsequently beaten, new waves of Viking invasions continued to plague Europe. In France, Ragnar Lodbrok and others besieged Paris. In the Mediterranean, Vikings wrecked town after town, sacking cities like Lisbon, Cadiz, Seville, Algeciras, Narbonna, Luna, Pisa, Fiesole, and Pamplona. They sacked the lucrative trading centre Durestad four times, incurring a devastating blow to central Europe. These *Northmen* (*Normans* in Frankish) seemed unstoppable.

However, they did not come just to plunder – they also came to settle. For some debated reason, the Norsemen embarked on a massive exodus out of their ancestral lands and swept across Europe to find better land. It was as if the events that led to the destruction of the Western Roman Empire (5th century A.D) were repeating themselves. Norsemen settled in Northern France (Normandy), Ireland, England, across the islands of the North Sea, in Frisia (in modern-day Germany), Sicily and across Eurasia. Here they carved in their own, new Kingdoms and Earldoms, and their Norse culture soon began influencing the European. Not only did languages exchange words and grammar rules, but traditions, art and worldviews merged.

The consequences of this process are undeniable: it led to the development of new ideas, cities, and even new countries. For instance, they were the founders of the Russian state – chief Ruric from Sweden founded the Rus kingdom in Novgorod, creating a dynasty that would rule all the plains from the Gulf of Finland to the Crimean Peninsula. It became the first organised authority on these wide plains and morphed into the Kievan-Rus Empire, marking the beginning of the history of Russia. On Ireland, the Norse Vikings founded the city of Dublin, developing a Viking-kingdom on Ireland that would rule the Irish sea for centuries.

So why did these Scandinavian Vikings travel so eagerly away to foreign territories? One theory holds that, in the face of enhanced trade with Europeans that produced more wealth, the population grew too large for the scarce Northern resources to support. Other claim there was a climate crisis in Scandinavia that worsened the prospects for food production, sending large sways of the population abroad. Many therefore migrated to foreign lands that had better farming opportunities. Regarding this, England was the jewel. The endless meadows of Britannia proved irresistible for the Norsemen. For centuries they would attempt, and achieve, to become Britannia's masters.

Norsemen valued the spirit of adventure. They were curious, and courageous. Any opportunity to defy the will of the waves by crossing the sea, find a distant land and engage in risky violence or mysterious exploration, ought to be taken. After all, they had only a limited time on Midgard –*Middle Earth* – before they had to enter the Spiritual domains of the gods. They longed for a good legacy to leave behind, hence the Norse proverb *"cattle die; comrades die; and you die too. But the words of a legacy lives forever."* They probably also sought the favour of the gods, who, in Norse religion, rewarded brave souls by various means. Upon a Norseman's death on the battlefield, the heroic warrior could be taken into *Valhalla* – the hall of the slain – where the dead would feast and wrestle with Odin until the end of time. Good farmers or merchants could enter the home of Frey, the god of wealth. Those who lived without accomplishments entered the dull world of *Hel* – where absolutely nothing worthy of remark happened. Thus, the afterlife would be spent with the gods until *Ragnarok* – the final battle and the end of the world. This mythology certainly promoted adventurism by means of war, exploration, or trade. It lacked a clear definition of good and evil, but still included a set of virtues that all Odin-worshipping Norsemen pursued. It emphasised ambition, boldness, strength, ingenuity and, sometimes, slyness.

The importance of gifts in Norse culture was also a strong incentive for the Vikings to go abroad. Their society evolved around it the giving and receiving of gifts. The man who had the largest collection of precious items was always followed. This chief would give gifts to his subjects, and in exchange, his subjects would give him loyalty. If a chief ran out of gifts, he would lose the loyalty of his men. This was expected. The more exotic the items were, the greater value they had. Chiefs who were considered good were therefore often described as "generous" and "eager givers" in saga literature. Throughout this book, we shall see many examples of the strategic value of riches.

Riches come either through trade, or through war. But the latter involved high costs and risks. For this reason, contrary to popular belief, the majority of Norsemen were traders, not Vikings (raiders). Their interest for the outside world greatly increased as the years passed by, and chiefs made sure to promote trade by investing in the development of trading cities, like Kaupang in Norway, Ribe and Hedeby in Denmark. A vast and impressive trading network emerged, which is yet another explanation for the rising wealth in Scandinavia. Controlling the trade routes towards the North became very important to the early kings in the region.

To the Far North, the Norsemen taxed Samis and orbited closely around the North Pole to trade with Inuits for their immensely valuable furs, hides and walrus ivory. Heading East, they crossed into Arkhangelsk and Siberia. On the long rivers of Dniepr and Volga, the Norsemen sailed to reach the Silk Road, Persia, Mesopotamia, Syria, the Holy Land, and the Byzantine world. Norse merchants were found in markets across Spain, Italy, France and even Africa. As we shall later discover, goods originating across the Atlantic also found their way into Scandinavian homes. This amazing, global network made the Norsemen truly international. Wealth like spices, silk, glass, metals, and Persian silver, flooded into the North like never before. The Norsemen in turn exported various furs and hides, wheat, honey, feathers, falcons, whalebone, walrus ivory, amber and slaves.

The Norsemen owed much of their success to their technologically advanced ships. The development of this new type of ship building, was one of the major contributions that made the Viking era possible. They exploited Europe's many rivers, using them as highways, easily fleeing from enemies in a quicker pace than other ships. The ingenious longboats were ideal for both open oceans and narrow rivers. They were slim and fast, but also robust. They were also light, so that Vikings could carry them on-land if necessary.

With such a static system of profitable trade routes, competition over the *control* of these routes naturally became a defining element. The Dniepr and Volga rivers were monitored by the Rus Norsemen, who taxed bypassing merchants and earned a fortune. The same tax was imposed on merchants travelling along the Norwegian coast, the *North Way*. Therefore, competition among the tribes and clans along the North Way was fierce. They constantly attempted to enlarge their naval territories at the expense of their neighbours, and thus reap great wealth in trade taxes and tariffs.

A similar situation could be found in Sweden, a land divided between the Swedes and Geats (or *Gots*). According to the old texts, Denmark had technically stood united under King Dan, but then split again. It was not reunited until 936, under Gorm the Old's reign. Gorm's son, Harald Gormsson "Bluetooth", led Denmark to become a mighty Norse kingdom, exerting great influence over Viken in Norway and clashing heavily with the Anglo-Saxons in England. However, Harald Bluetooth also converted from Norse paganism to Christianity – a remarkable moment.

Christendom and Norse paganism had stayed at war since Charlemagne's invasion of pagan Saxony (772 AD). Despite these hostilities, the Christian faith itself gained traction in Scandinavian societies. The brave missionary St. Ansgar – *the Saint of the North* – founded Christian communities in Denmark and Sweden. Priests and monks from Ireland spread the Scriptures along the North Way. At first, the Norsemen tolerated Christian activity. Some agreed to be baptised - but it soon became complicated. As Christianity began spreading rapidly through the work of the missionaries, Norsemen felt threatened. It was more than a question of religion – this was a question of culture and worldviews. A culture clash was enveloping.

Kingdom of Vikings

It is in this environment that we begin our journey through the thrilling lives of the men who forged the Kingdom of Norway. It was with this worldview, this history of war and conflict, this knowledge of seas and lands, that the first Viking chiefs originated the idea of a united country and would give their lives to realise it. The implications this had on Northern Europe (and beyond) were substantial – but furthermore, the life stories it contained are some of the most exciting and dramatic ever recorded. We may open this book by thanking Snorri Sturlason, the old Icelandic warrior and author, along with all the countless other saga-writers, for putting these tales on paper and saving them from being lost in memory or myth. Thanks to them, we now have a rich set of biographies at our disposal, including intriguing details and vivid stories.

This book intends to re-tell these stories but through a modern lens. It maintains the lively storytelling of Snorri Sturlason and the old saga-writers, but it adjusts its narrative to accommodate for modern scholarship. The chapters, each covering a life of a significant figure, describe the life events in a concise and easily understandable manner, while keeping an entertaining tone to make room for the reader's imagination. Sometimes, dramatic moments are often prefaced by *"according to the sagas"* or similar, to specify that the exact quotes or details of the episode are directly taken from the sagas. These short biographies are therefore sourced *both* directly from medieval sagas and contemporary ballads (of witnesses), *and* from assessments by modern historians. The bibliography of each chapter can be found at the end of the book.

Can we trust the sagas and medieval sources? They were written in the 1200s, so they covered events ranging 300 to 50 years before them. Despite being far closer to the original events than we are today, many modern scholars have chosen a very sceptic approach to the sagas. Academic scrutiny is, of course, important, but the sagas have arguably been often proven to be quite accurate. At least, the brunt of their narrative is usually confirmed as legitimate.

We must remember the efficiency of oral lore – the style the Vikings used to record and transmit information and stories. Witnesses of notable events would use a common syntax and rhyming scheme to record the events in *kvad* – poems or songs. This is known as *skaldic poetry*. The rhyming made them easy to remember, and they were sung or repeated on many occasions to spread them to others. It was important they were sung *correctly*, to maintain the details of the lyrics. This is why each Viking King or lord usually had several *court poets* or *skalds* in his staff, so they could begin recording what they witnessed. Some would, of course, have a favourable bias to their employer – but many *skalds* also composed poems to *insult* or criticise chiefs and Kings. Snorri and other saga-writers used these songs and ballads, but also reviewed other manuscripts written before their time, and inspected local knowledge and memories. The combination of all this produced the sagas. Let us hear how Snorri Sturlason himself defends his texts in his own preface:

"In this book I have had old stories written down, as I have heard them told by intelligent people, concerning chiefs who have held dominion in the northern countries... Some of this is found in ancient family registers, in which the pedigrees of kings and other personages of high birth are reckoned up, and part is written down after old songs and ballads which our forefathers had for their amusement. Now, although we cannot just say what truth there may be in these, yet we have the certainty that old and wise men held them to be true."

This book – *Kingdom of Vikings* - consists of two parts. The first covers the founders and first Kings of Norway under the Viking Age (795 to 1066). The second covers the *forging* and maturation of the Kingdom as it enters the Middle Ages (1066 to 1280) – an era of unprecedented growth in the Kingdom, but also brutal conflict and notorious civil wars. The Kingdom of Norway was borne out of the Viking Age. It was hammered together by Viking Kings. This legacy was very dear to all the Kings who took upon themselves the throne of Norway. All strived to honour it, and many battled to attain it. This is the heart of the historic Norwegian culture and identity. It was, and is, a nation carved out in the Far North of the world. Norway is the Kingdom of Vikings.

Terms

Norse religion, or **Asatru**, was the old Germanic religion of the Northmen.

Thing was a governing assembly among the landowners (farmers) in a respective region. They discussed local issues, passed laws, resolved quarrels and, if they wanted, nominated a king or chieftain. The matters were decided upon by voting. The thing is at the *heart* of Norse society. It was the only legislative institution. Chiefs and kings were always at the mercy of the will of the thing. The common English word for *thing* – an object or item – derives from this word. In Norway, thing assemblies ruled over a particular jurisdiction. Its laws were effective only in that jurisdiction, and only landowners from that jurisdiction could attend and vote in the thing. This gave Norway several law-codes in various provinces across the country, most prominently the *Gula*-code of *Gulathing* and *Frosta*-code of *Frostathing* (see *Magnus the Law-Mender*).

The **Odel Tax** was an inheritance tax on private property. It was enacted in Norway by King Harald Fairhair, but was extremely controversial. Farmers were free men and their land were wholly theirs, so the idea that a King could require tax on inherited property, was a cultural insult. Many regents faced stubborn resistance from farmers because of the infamous odel tax.

Blot was a Norse pagan ritual. Animals, and sometimes humans, were sacrificed and their blood were sprinkled on the statues of the gods, on the pagan temple walls and on the spectators. The session was then followed by a feast where the participants would eat the boiled meat and drink ale.

Blood feud or **blood vengeance** was a Norse tradition that effectively permitted murder if it was done to avenge the murder of a relative, close friend or slave. The blood feud was inherited, so it could often lead to generations of vindictive strife.

Viking is a Norse term for raider. It could also be a verb, i.e, *to go viking* meaning to go raiding. It is therefore *not* a general term for an ethnicity, people or tribe. The peoples of Scandinavia in the dark ages were instead known as the *Northmen,* but their habit of raiding and plundering also closely associated them as *Vikings* in the eyes of their victims.

✳ Terms ✳

In modern view, the term *Viking* has been popularised and often used interchangeably as a term for the distinct culture and people of the Scandinavians – also giving the era its name, the Viking age.

Northman was the more common European expression for the peoples of Scandinavia. It may originate in Frankia (modern France) as *Northmann,* later Latinised as *Norman.* Modern Norwegians are still called this in Scandinavian languages (Norwegian: *Nordmann*; Danish: *Nordman*d; Swedish: *Norrman* or *Norska*)

Norse is a term used for any person or community that adheres to the old Germanic Norse culture and belief system (*Asatru*). **Norseman** is a more anthropologic and modern term. It was first coined in the 19th century, attributed to the peoples of Scandinavia that trace their origins to the original Norse Germanic culture. This book uses Norseman quite often as a term for the Norse peoples of Scandinavia.

Danegeld was a tax paid to guarantee safety from Viking invasions or plundering. It was usually paid by the victims to the aggressors, the Vikings. It directly translates *Dane-debt.*

Danelaw was any foreign area subject to Norse laws, but was particularly pertaining to England. It originated after English-King Alfred the Great's treaty with the Vikings in 878, where a body of Danish law was still in force in large sways of England. The term was then often used to mark where the Vikings held dominion (i.e, where their law ruled) in England.

A **Seer** was a Norse fortune teller or wizard but could also be used for anyone who was considered prophetically gifted. For example, Vikings could call a Christian prophet as a *seer.*

Earl of Lade (Jarl of Lade or Ladejarl) was perhaps the most powerful title in Norse-Viking Norway, founded initially by Haakon Grotgardsson. It held Earldom over Trondelag and neighbouring territories, thereby controlling the most populous and mighty people. The key to control over Norway lay in Trondelag, and therefore lay with the Earl of Lade. Only with the rise of feudalist structures would the might of the Earl of Lade decline. These Earls also followed a dynastic lineage as the title was inherited from father to son.

Maps

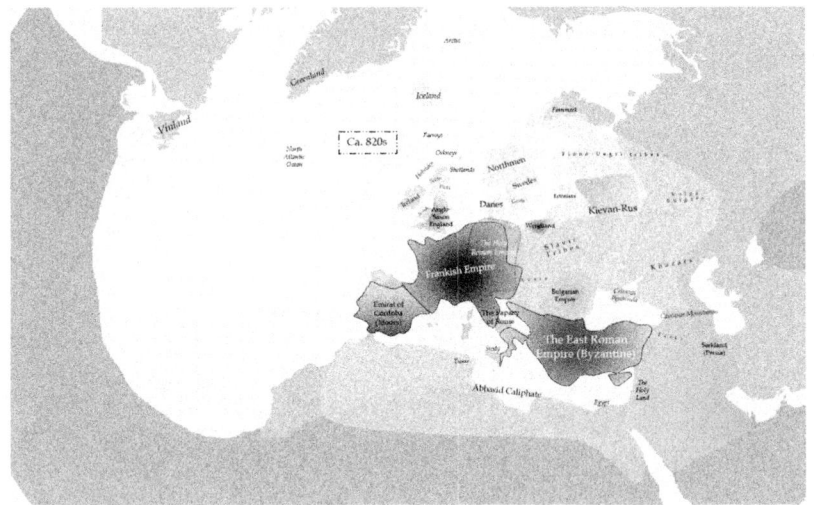

Map 1: "Middle Earth" around the 820s.

✶ Maps ✶

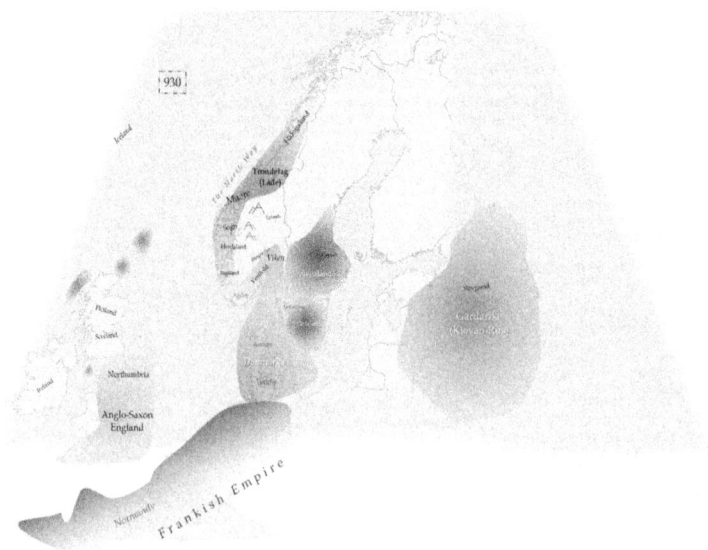

Map 2: Northern Europe anno 930.

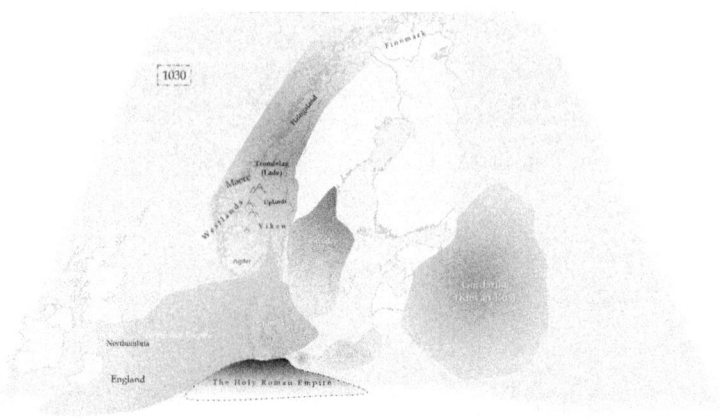

Map 3: Northern Europe around 1030 (before St. Olaf's fall)

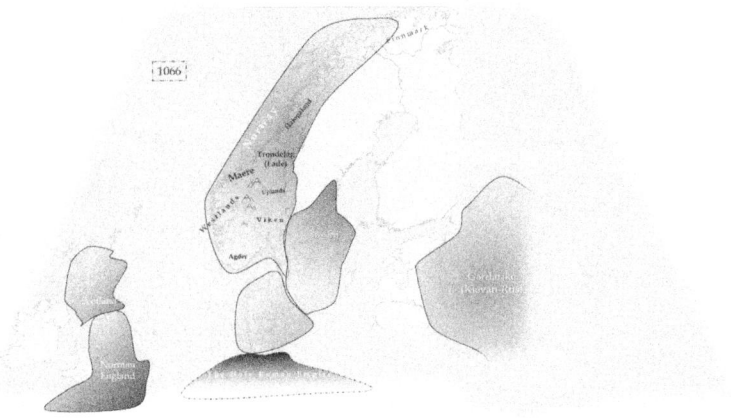

Map 4: Northern Europe anno 1066 (after the Battle of Hastings).

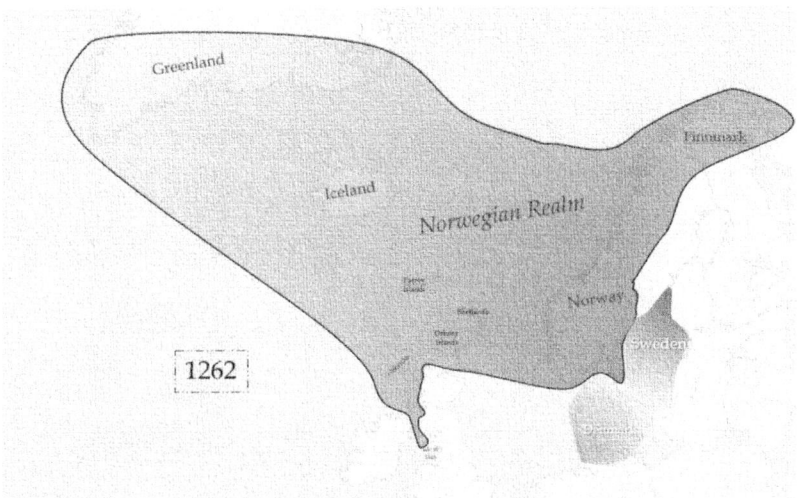

Map 5: The Norwegian Realm under King Haakon Haakonsson.

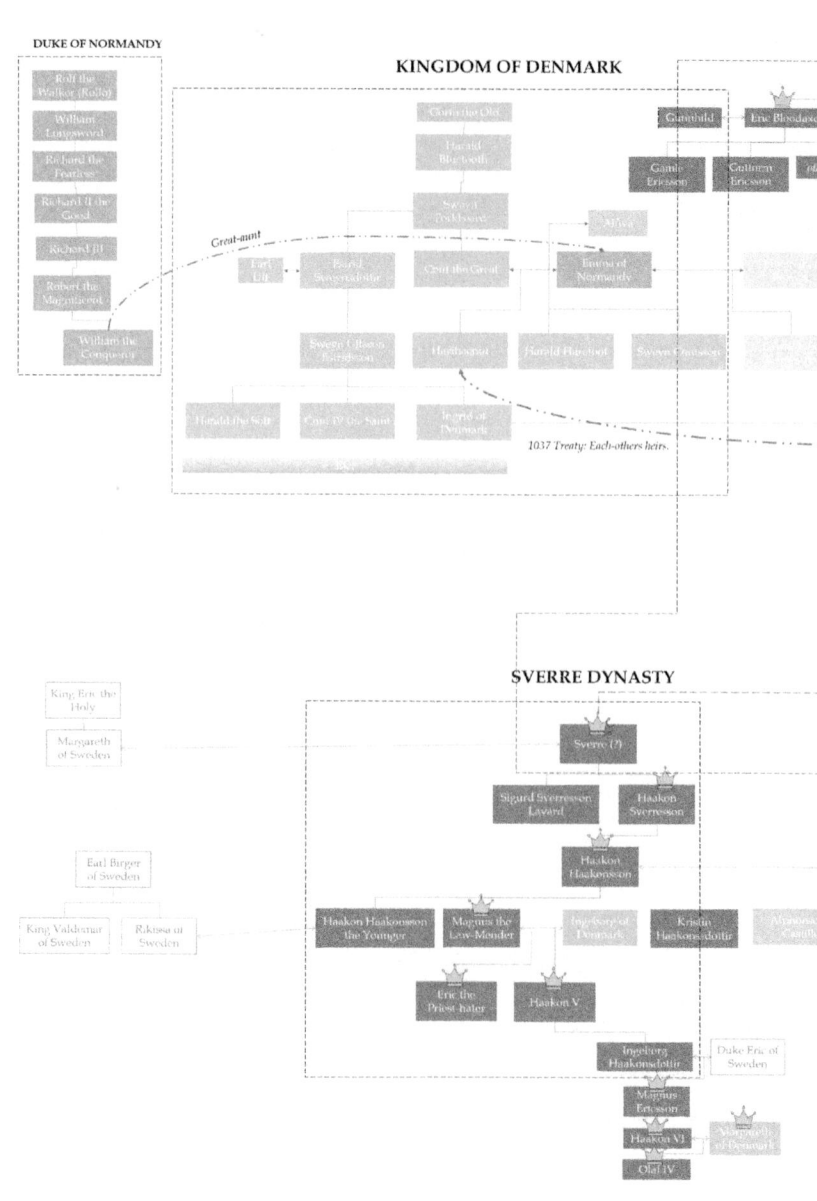

DUKE OF NORMANDY

- Rolf the Walker (Rollo)
- William Longsword
- Richard the Fearless
- Richard II the Good
- Richard III
- Robert the Magnificent
- William the Conqueror

KINGDOM OF DENMARK

- Gorm the Old
- Harald Bluetooth
- Sweyn I Forkbeard

Gunnhild — Eric Bloodaxe

- Gamle Ericson
- Guthrum Ericson
- others

Great-aunt

- Earl Ulf
- Estrid Svensdottir
- Cnut the Great
- Emma of Normandy
- Alfiva

- Sweyn II the Estridson
- Harthacnut
- Harald Harefoot
- Sweyn Cnutsson

- Harald the Soft
- Cnut IV the Saint
- Ingrid of Denmark

1037 Treaty: Each-others heirs.

SVERRE DYNASTY

- King Eric the Holy
- Margareth of Sweden

- Sverre (?)

- Sigurd Sverresson Lavard
- Haakon Sverresson

- Haakon Haakonsson

- Earl Birger of Sweden

- King Valdemar of Sweden
- Rikissa of Sweden

- Haakon Haakonsson the Younger
- Magnus the Law-Mender
- Inge-Gerd of Denmark
- Kristin Haakonsdottir
- Afamous ? Family

- Eric the Priest-hater
- Haakon V

- Ingeborg Haakonsdottir
- Duke Eric of Sweden

- Magnus Ericson
- Haakon VI
- Margareth of Denmark
- Olaf IV

Family Trees

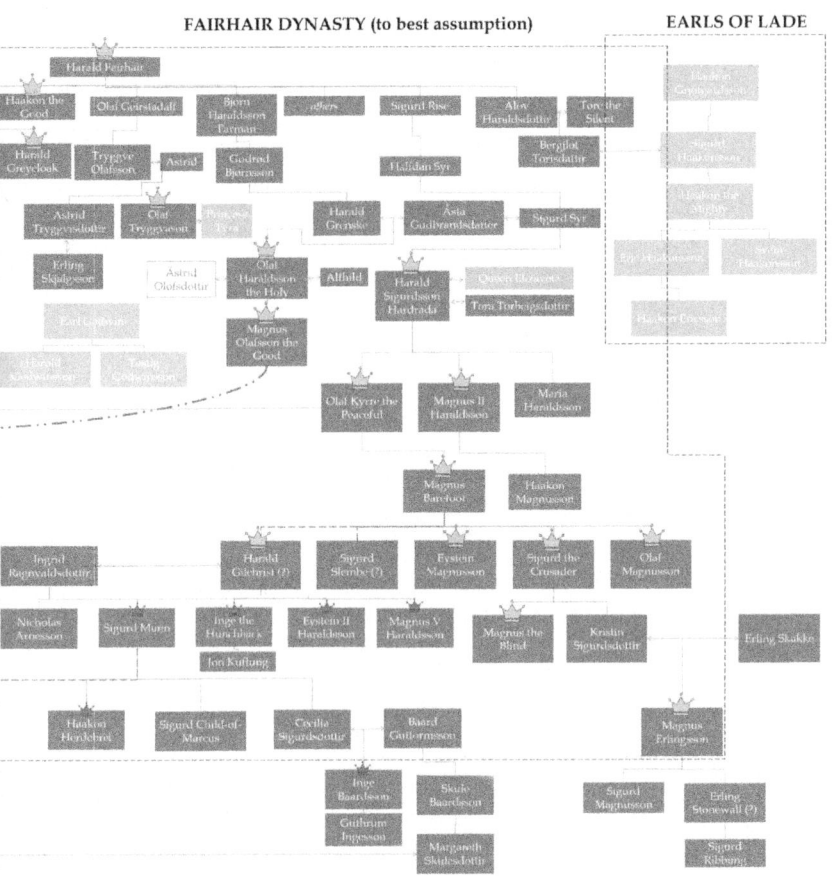

PART I

Harald Fairhair

850 – 933

"I make the solemn vow, and take God to witness, who made me and rules over all things, that never shall I clip or comb my hair until I have subdued the whole of Norway, with tax and duties, and domains; or if not, have died in the attempt."

– Harald Fairhair, from *Harald Fairhair's Saga.*

Harald Fairhair can be considered Norway's original founding father. Under the leadership of this man, all significant lands of Norway were, for the first time, assembled under one High King. He was a man of uncompromising will, relentless fighting spirit and a grand vision for himself and his dynasty. His fierce resistance to foreign interventions, clever tactics and strategies, introduced the Kingdom of Norway to the world. His story is dramatic, brutal, and important.

Royal beginnings

Harald Halfdansson was born in 850. He was the son of the ambitious king Halfdan of Vestfold, who again traced his lineage back to Odin, making him a member of the mythical Yngling dynasty. Halfdan, called *the Black,* had spent most of his years hammering together a larger kingdom through diplomacy, intermarriage, intrigue, and brutal warfare. His dominion now included the lands of Vestfold, Ringerike and Sogn.

Halfdan sent his son to be raised in Sogn. The little prince grew up in the beautiful fjords and was taught and tutored by old warriors and lawyers. He soon showed remarkable intellect, fighting skill and a deep ambition to match or exceed his father's achievements. His chance to do so came sometime in the 860s, when Halfdan the Black fell into a frozen river and died. As the crown passed safely on to Harald, he prepared how to lead the realm to an even greater extent.

Harald's kingdom was positioned very well strategically. There was active trade along the Norwegian coast, with hundreds of trading ships sailing from Europe and to the far North of Norway. Harald's kingdom controlled a good portion of the North Way, granting him substantial amounts of funds through taxes and tariffs. In addition, his holdings in Vestfold gave him access to the trading cities of Viken. Among the most lucrative of these were the city of Kaupang, meaning "*marketplace*". This gave him a solid foundation for raising an army.

So, what were the ultimate goals of the young king? Like historian Kim Hjardar points out, Harald desired to control all the petty kingdoms along the North Way to secure the entire trade route for himself. If he did this, he would be rich beyond measure, and more powerful than anyone who lived before him. A respected seat by the gods in Valhalla would be guaranteed.

But of course, there is also a different version of how this plan came to mind. Saga legends have it that when Harald romantically proposed to the beautiful Gyda, a woman Harald deeply admired, Gyda rejected him, declaring instead that she would only marry him if he was the ruler of all Norway. Harald took these words to heart and swore to never cut his hair until all of Norway was beneath him. Due to this, Harald won his epithet "*Hårfagre*", or as it is in English, "Fairhair".

Such a conquest was not an easy task. Harald spent years together with his father's advisors, landowners, and veterans, planning how to effectively subdue all the petty kingdoms. Thanks to Halfdan the Black's campaigns, Harald's advisors were experienced in the art of war. Most notable was the veteran Guttorm, a fine warrior and clever tactician. They advised to resume Halfdan's *whip and carrot* strategy. It was straightforward: reward vassals with greater riches than they could ever dream of as independent petty kings; punish opposition with brutal force.

Harald already had substantial funds raised through commercial taxation along his share of the North Way, but to raise a truly invincible force, he needed more. The solution was a special tax on inherited wealth – the *Odel tax* (see Terms). This was incredibly controversial because Norwegian farmers had always enjoyed the right to private property with no infringement or tax by petty kings. This was something that made Norway quite special – its farmers were fully free and independent. Harald however, sought to change that. Inheritance tax on private properties would create another, lucrative source of income that would enable him to finance a vast, unchallengeable army.

The Conquest of Norway

Ironically, Haralds offensive campaigns began on the defence. Neighbouring kings sought to overthrow him, as they assumed him to be a naive ruler. King Gandalf of

Upland was the first to strike. He sent his son Hake to invade Vestfold by land, while Gandalf would simultaneously invade by sea. Harald was too young to lead the army, so he sent the warrior Guttorm in his stead.

Guttorm moved with speed. He first defeated Hake in a valley, and managed to have Hake killed (today, this very valley is called *Hakedal,* meaning Hake's valley). Guttorm then turned to the coast to stop Gandalf. Gandalf did not pass. Guttorm routed his army. Already from these early engagements, it was clear that Harald had quite an efficient military machine ready to be used.

Around 863 AD, Harald Fairhair reached maturity and began his first campaign. As Gandalf and Hake had been defeated, the Uplands were already subdued. Harald therefore marched further North to reach Trondelag, the most prosperous and powerful region in Norway. While marching, he terrorised all his opponents with ruthless brutality, torching households, barns, and farms if necessary. He quickly gained a feared reputation and many Norwegians began surrendering at once. Those who surrendered were absorbed into Harald's ever-expanding army. By the time he reached Trondelag, he was already known as an unforgiving warlord.

This provided much leverage to Harald when he negotiated with Earl Haakon Grjotgardsson, the Earl of Lade (see Terms) and effective lord of Trondelag. Earl Haakon feared Harald and preferred peace over war. They reached an agreement that ultimately made Earl Haakon a vassal of Harald, in exchange for peace. To seal the deal, Harald agreed to marry Haakon's daughter.

This was a stunning success. Trondelag held plenty of tough fighters, skilled traders, and seafarers – and Harald had secured access to all these vital resources without having fought a single battle. He ordered the construction of a warfleet, and as soon as it was ready, began sailing south along the coastline to subdue the other rivalling petty kingdoms and earldoms, one by one.

It was here that Harald met fierce resistance for the first time. The maritime seakings of Maere (name etymology is possibly linked to *Mare,* the Latin word for *sea*) formed a coalition to stop him. They were probably very disappointed by the lack of resistance from the Tronds. As Harald neared the important maritime pass by the island of Ed (*Edøy* – an ancient seat for maritime kings and Vikings in Norway), the seakings of Maere ambushed him with a major warfleet.

This became the massive naval battle of Solskjel. The first engagement was inconclusive, but a second engagement soon followed. After hours of gruelling combat, Harald stood victorious. The seakings were forced to flee or surrender. Many fled to the various Norse islands of the North Sea, where they founded new kingdoms, chiefdoms and earldoms, and planned to continue the fight against Harald.

Abruptly, the King heard news that Swedes and Geats, two strong peoples in modern-day Sweden, threatened his lands to the East. King Eric of Swealand had begun taxing farmers on Harald's lands in Ostfold. There was also reason to believe that the Swedes planned to invade Viken and Vestfold – such an incursion would strike at the heart of Harald's realm. With no time to waste, Harald galloped to the East. He slaughtered everyone in opposition and razed entire villages to the ground. King Eric of Swealand was assassinated and many battles were fought until the Swedes were beaten into a ceasefire. Next, Harald stormed the lands of modern-day Ostfold, drove the Geats away and marched deep into enemy territory, reaching the Göta River. Here he was stopped, but it suited him well. The river served as a natural border to the Geats and Swedes. Harald appointed the old Guttorm as commander of the Eastern territories before returning West to resume his campaign along the coast.

Battle of Hafrsfjord[4]

What remained was for the King to subdue the South. This would be, by far, his greatest challenge yet. In the South of Norway, the Danes controlled many settlements and exerted much influence. In addition, rich and powerful kingdoms reigned there, with political connections across the North Sea. When Harald opened his Southern campaign, the Norwegian kingdoms came together and formed a united front against him. Harald had upset the balance of power in Norway and threatened their sovereignties. They called upon their Western allies for help. Denmark naturally joined them, as they feared that Norway under Harald's absolute rule could create a new, Northern power that would challenge the Danish hegemony and ruin their commercial interests in Viken. Norsemen and Viking raiders from Ireland, the British Isles, the North Sea isles and the rest of Norway, joined forces to stop Harald once and for all. The final war for Norway was about to enfold.

Harald's spies reported to him all the movements of the enormous enemy alliance – but the King remained calm and collected. To him, this was a good opportunity to lure all his enemies into *one* defining battle that would decide the fate of Norway.

His next move is quite telling: he established his military headquarters in Hafrsfjord in Rogaland, part of Southern Norway. This was a fjord consisting of a narrow pass

---

[4] Can be pronounced *Haf-s-fyord*

leading to a wide lake-like area. It was a brilliant trap. With such a narrow passage for entry and exit, he could easily trap the enemy fleet by closing off the narrowest point.

He then began to raid and campaign around Rogaland, seizing major farms and executing those who revolted. This clearly provoked the enemy alliance, causing them to speed up their mobilisation efforts and, likely, act out of anger and desperation rather than strategic calculations. Time was on Harald's side, not theirs. By 872, they had seen enough of Harald's pillaging and went to action, attacking him head on in Hafrsfjord.

Despite them falling into Harald's plan, it was also clear to him and his men that it was a very risky gamble. If he lost at Hafrsfjord, he would lose his military headquarter in the South, and therefore lose his warfleet (which would also be trapped in the fjord), supplies and resources, and his winning momentum. This sign of weakness could trigger a lethal counter-reaction against him, untying the realm he had fought so hard to forge.

The enemy alliance entered the fjord bay and took up position. The bay now held 10 000 vicious Northmen, all embarked in longboats and ready for battle. Harald decided to deploy his ships in a tight formation so that they resembled one, massive floating fortress. He even ordered his men to tie their ships together with ropes. By contrast, the coalition fleet was loosely coordinated, consisting instead of a score of individual longships.

The battle began as the alliance charged towards Harald's centre, bound on boarding Harald's royal flagship and kill the King. Harald, in turn, ordered his elite *Dragonships* to detach from the main fleet and pursue the flagships of the alliance. The King then revealed his most dangerous weapon yet: the notorious berserkers. These tall warriors were infamous for their explosive and almost uncontrollable style of fighting. They threw anchors onto enemy ships, jumped aboard and hurled their axes at the men, inflicting heavy casualties on the alliance and spreading fear among the ranks. After just two hours, the allied fleet broke down and ships began to retreat – soon, it led to an all-out rout. Harald had the narrow exit pass closed and, just as he had planned, had trapped a demoralised, fleeing enemy. Disaster ensued: hundreds drowned, while others reached land and began to flee by foot. Most of the enemy commanders were slain or executed. Harald Fairhair had won a crushing victory.

The petty kingdoms of Southern Norway stood leaderless and succumbed to Harald's authority. Danish forces and settlements were chased out of the country. Harald then purged all his enemies, their families, and other suspects. Many were executed, exiled, or fled the country. Enemies of Harald were scattered across the Northern Europe. For the first time in history, all of the significant lands along the North Way were under the rule of one, single king. Harald Fairhair was the master of the North Way.

Discovery of Iceland and Rollo's Normandy

The consequences of Fairhair's conquests had far-reaching effects. His warpath had stirred unrest in many communities, causing its inhabitants to flee from Norway and seek a new home. This triggered a re-settling of various Norwegian chiefs and petty kings, and, through a remarkable sequence of events, would change Europe forever.

One of such cases was Ingolf Arnarson, who fled his hometown in this period. Instead of seeking re-settlement in England or the North Sea isles, Ingolf had decided to gamble on an uncharted island to the Far North-West. Rumours spread that there was a major island there, virtually empty, with good and wide landscapes for farming. The talented Norwegian shipbuilder Floki Vildergarson had already investigated it and found it, though he later left the island and called it *Iceland* (perhaps out of spite). Ingolf took his brother Hjorleif, a talented navigator, his family and a team of seamen and set out to settle the mysterious place.

After weeks of sailing the wild sea, they were relieved to finally spot land. Ingolf threw his high seat pillars (symbols of a chief) overboard and swore to found a settlement wherever they landed. They were later found in a small bay, where Ingolf and Hjorleif founded Reykjavik – Iceland's current capital. News of his successful voyage and settlement spread fast, and soon, hundreds of Norwegians came to start a new life. Iceland would become a "second Norway", but develop its own, distinct Norse culture.

Another remarkable consequence of Fairhair's campaigns were the case of Rolf (or *Hrolf*) the Walker[5]. According to the sagas, Rolf was a seaking of Maere, but after a conflict with Fairhair, he was exiled. He raided across Britannia, gathering a large Viking force, before joining a coalition of Vikings to besiege Paris in 876. The Franks paid the Vikings away in silver – but they also made an unprecedented offer to Rolf. They would give Rolf the wealthy town of Rouen in Northern Frankia and make him Duke. In return, he had to convert to Christianity, become a vassal, and serve the King of Frankia to stop all further Viking plundering. He accepted.

Rolf married the daughter of the Duke of Rennes and encouraged his subordinates to take local wives. He learned Frankish, Christianity, and his name was Latinised *Rollo*. He soon gained influence in the Frankish court, and when a usurper tried to coup the throne, Rollo took an active role in defending the legitimate lineage of Frankish Kings. For this, he was given more fiefs and lands in Northern Frankia.

---

[5] Hrolf was called *the Walker* because he was apparently so big that a horse couldn't carry him.

Soon, scores of Norwegians and Danes came to settle in his lands, and the entire region became known as *Normandy* – the land of the Northmen (or *Normans,* a Latinised version of *Northman*). The Normans became an elite fighting force for Christian Europe. They drove out Muslim invaders from Southern Italy and Sicily, and served as the elite units under the First Crusade in 1096. Many French towns were also named after these Scandinavians. Hauteville, for instance, is named after the fierce Viking Hjallt, one of Rollo's closest companions.

Rollo lived at peace until his death in 927 – but the dynasty he founded would go on to do great deeds. His great-great grandson was William the Conqueror, the man who won the throne of England in 1066 and, in turn, founded the British monarchy. Who would have known that the current-day British monarchy would trace its lineage back to this Norwegian chief from Maere?

Reign and Dynasty

By sometime in the 870s, the entire North Way, from Agder to Trondelag, was finally laid under Harald Fairhair's authority[6]. Funds streamed into his treasury from all the tax profits he made. The inheritance tax made him unpopular among the farmers, but it contributed substantially in filling up his coffers. Besides, he ruled by an iron fist. His royal seat was almost impossible to tear down.

Though Harald would experience some inner uprisings, his brutal reprisals generally ensured that the people remained obedient. As we have seen, many chiefs instead chose to flee the country. Farmers who felt deeply provoked by Harald's encroachment on their private property sailed for the newly discovered Iceland and settled there. Former chiefs, earls and even kings would also relocate across the North Sea – Shetlands, Faeroys, Orkneys, Hebrides. Here they plotted to recapture their traditional holdings and defeat Fairhair, but their plans never materialised.

Harald Fairhair spent the years post-Hafrsfjord to consolidate his authority and build his own dynasty. As historian Dr. Torgrim Titlestad pointed out[7] Harald's new regime seemed to take form as a confederation. On top of the hierarchy was Harald Fairhair, the High King of Norway. Below him were regional kings, appointed directly by Harald. Dr. Titlestad outlined two essential ways by which Harald would ensure the

---

[6] There is uncertainty whether Harald actually exerted authority as far North as Hålogaland.

[7] See bibliography

loyalty of these regional kings: he used notorious Vikings as a form of secret police to arrest, threaten or assassinate political enemies; and he inter-married with other royal families. The latter helped him produce quite a sizeable dynasty. He allegedly had 10 wives and 11 concubines and produced 20 children, where the majority were sons.

However, King Harald also needed to address his foreign relations. The Swedes were held in place to the East and the Danes had been successfully driven away from Viken, but only temporarily. Knowing that the Danes were formidable fighters, Harald sought for a way to ensure that the Danes would not attack Viken again. He therefore married Ragnhild, daughter of the Danish king of Jylland. The marriage was useful, but not enough. He needed some geopolitical alliance to warn off Danish invasions.

The Anglo-Saxons were the perfect solution. King Athelstan of England had defeated the Danes but was looking for allies to help him protect his shores. Harald and Athelstan opened negotiations and it did not take long before they agreed to become allies. Together, they could defend each other's realms against Danish plundering. The political climate in the North Sea was changing fast.

Harald went on to subdue the Norse islands of the North Sea, allegedly because he wanted to destroy the Vikings there, who often ravaged his trade routes and towns. The King first took Shetland and *"slew all the Vikings,"* then headed to the Orkneys and did the same. When he stormed through the Hebrides, he cleared the area of Vikings, but also *went on Viking* himself, plundering Scottish fiefs to pay and satisfy his army. By the time he reached the Isle of Man, all its inhabitants had already fled to mainland Scotland, fearing this intruder. Harald Fairhair established Earldoms over these islands and appointed his own, trusted men to govern them, before returning home. He had thus secured total domination and humiliated all his enemies.

The remainder of Harald Fairhair's life was spent on consolidating his royal power, tutoring his sons and handling his wives. His sons were carefully trained in the art of warfare, the Norse religion and leadership. He bestowed kingship on many of them and assigned lands for them to administer as his representatives. However, his sons soon gained a reputation of abusing their power. Word spread that they mistreated commoners, seized private land and taxed their subjects harshly. Probably hoping that at least one of his sons would become a responsible ruler, the High King Harald sent his intelligent young son Haakon to King Athelstan, hoping that the boy could be raised at the Anglo-Saxon court and be tutored by scholars. Athelstan accepted this request and embraced the child as his own.

In 930, Harald was too old to reign and was aided by his most able son at the time, Eric Haraldsson. In 933 however, High King Harald Fairhair was moments away from dying. He declared his "favourite son" Eric to succeed him, before drawing his last breath and entering the afterlife. He became 83 years old – very old for the time.

Legacy

To Norwegians, Harald Fairhair is a universally known character. His vision, ambition and willpower made the concept of a united Norway a reality. Whether intentional or not, he helped spark the idea of Norwegian unity – an idea that would linger on and inspire generations of Vikings.

To Harald, there existed only friends or foes. Friends were rewarded with gifts and privileges. Enemies were punished ruthlessly with plunder, torture, execution, or exile. In this way, he attracted those who were willing to serve him, and intimidated those who would not. *"Only a fool trusts a smile, for treachery is always hidden,"* says a Norse proverb. Harald ensured himself that he had no treacherous, hidden enemies in his court.

This meant that the myriads of Norwegian chiefs and petty kings who denied Harald, fled, and as we have seen, left a lasting impact on Europe. The scale and terrible force at which Harald Fairhair stormed through the North Way therefore had major implications in both the Scandinavian world – and the world itself.

Most importantly, as he was the first unifier of all the Norse lands on the North Way, he became the founder of the *Fairhair Dynasty*. This was the *only* bloodline with rightful claims to the throne of the nascent Kingdom. Its lineage stretched back to Odin – there could be no other King over the North Way than a descendant of Odin and Harald Fairhair. Thus, this dynasty would be at the heart of the struggle for power and dominion over the North Way.

Haakon the Good

920 - 961

While Harald Fairhair was the man who assembled the North Way, Haakon the Good was the man who turned it into a functioning and more cohesive Kingdom. Overshadowed by some later Norwegian titans, Haakon is often found left in the dark by modern historians. However, this is a *severe* mistake. His organisational skills, administrative reforms and military leadership turned the North Way into a prosperous and robust Kingdom, preventing it from falling apart, which it was at the brink of doing. He laid the foundations of the nation. This makes Haakon, by far, one of the most important leaders in Norwegian history.

A Child of Athelstan

Eric, the son of Harald Fairhair, successfully inherited the throne of the Kingdom – but had not yet secured it. When he began suspecting his brothers of jealousy, he began a purge against his own family. One by one, his own brothers were hunted down and executed. It was a gruesome affair, even by the standards of the Vikings. For this, Eric was given the eerie epithet *Bloodaxe*.

Next, Eric increased taxes and seized land to finance a large scale Viking expedition into the Baltic. He raided notoriously along the coast. However, the people of the North Way were deeply dismayed by this new, erratic High King. A rebellion was looming.

Eric had, of course, one remaining brother who could still pose a threat: Haakon Haraldsson, who they called *Adelsteinsfostre* (Child of Athelstan), was safe and sound in England. When he heard of his fathers passing and Erics assumption to the throne, he was around 15 years old, and very well educated. Haakon was appeared quite intelligent and talented. The Anglo-Saxons called him *Haakon the Trustworthy* and *Haakon the Victorious*. He had adopted Christianity, learned politics, state management, the art of war, public speaking, and negotiation. He also seemed to be ambitious, for as soon as he heard of Eric's succession, he stormed to the English court and asked King Athelstan to allow him to return to his ancestral homeland and claim the throne. Athelstan, who loved his adopted son dearly, blessed his journey and gave him ships, soldiers, priests, and scholars. In addition, he bestowed to Haakon one of his finest longswords: *the Kvernbit.* With it, Haakon would conquer.

The teenage prince first landed in Trondelag, which he rightfully understood as the key to the power of the North Way. As mentioned before, the key to power lay in Trondelag.

Fortunately for Haakon, the Tronds were already preparing for an open rebellion against High King Eric. Haakon immediately arranged a meeting with their leader, Earl Sigurd of Lade, and forged an alliance with him. Together, they organised a *thing* (see Terms) to speak to the chiefs and landowners of Trondelag. Having gathered all men of significance across the land, Haakon rose to speak. He was an eloquent orator and promised to abolish the infamous odel tax, restore the right of inheritance for landowners and bring justice to the monarchy. By doing this, Haakon promised to ensure the integrity of the earls and landowners and uphold the ancient Nordic tradition: The King was to serve the people, not the other way around. The Tronds applauded him at an instant and hailed him as their new High King.

With the backing of the mighty Tronds, Haakon had already become a significant threat to his brother. He continued touring the country, holding things in various communities, offering gracious terms and generous gifts, and winning political support at every assembly. In this way, Haakon secured popular support and therefore outmanoeuvred his brother politically without a single battle. He also helped prevent a total disintegration of the nascent Kingdom of the North Way, as an open civil war could cause irreversible fragmentation.

Eric Bloodaxe had already lost. His peers deserted him and only a handful of his former friends still supported him. One day, he received an invitation by King Athelstan to come to England and serve as Earl of Northumbria instead. It was an honourable exit that provided Eric with a "way out". He knew that war with his brother was futile without popular support, so he packed up and sailed for London. With his abdication, Haakon had won the Kingdom.

Although not investigated much by scholars, it is very likely that this was a planned sequence of events between Haakon and Athelstan. They had planned what to offer the Tronds and the people, what routes to take and what things to call – and that Athelstan would, at the right moment, provide Eric with a convenient exit to England. It certainly seems very well-coordinated.

The Fate of Eric Bloodaxe

Eric Bloodaxe did not disappear from history when leaving Norway. In fact, his career had just begun. He met with King Athelstan and was given a similar offer to the one Rollo had once been given: he was to be given York but had to convert to Christianity and defend Northumbria from Viking raids. This would be a perfect solution for Athelstan. Ever since he captured Northumbria from the Vikings in 944, Norsemen kept returning to reclaim it. This was not surprising: Vikings had controlled Northumbria since the 870s, and the population was largely Norse. This was why it also was so convenient for Athelstan to have Eric Bloodaxe, a Norseman, govern it. Given his talents in warfare, plus his relation to Harald Fairhair, Athelstan's former ally, Eric Bloodaxe was the perfect man for the task.

Earl Eric enjoyed a lavish lifestyle as he sat in York. He defended it well against both the state's enemies and his personal ones. For instance, his legendary enmity with the berserker and poet Egill Skallagrimsson was finally settled at the court of York, where Egill surrendered by performing a *kvad* (poem) that praised Eric's legacy.

Maybe such words helped feed Eric's ego and ambition, because in 948 he forcefully expelled the Anglo-Saxons from York and took the title *King of York,* thus reviving the independent Norse Kingdom of Northumbria. He experienced invasions from the Dublin-Vikings, but also from the new King of the English, Edmund. In 954, while marching North, Eric was ambushed near Stainmore and killed in action. His death gave way for the Anglo-Saxons to retake Northumbria, therefore ending Norse rulership over York once and for all. Nonetheless, Eric Bloodaxe's legacy was celebrated among his Norse contemporaries, who hailed him as a man who embodied the Norse virtues - a man worthy of Valhalla.

Eric was survived by his wife, Queen Gunnhild, six sons and two daughters. Queen Gunnhild was equally, if not more, ambitious than Eric. News of her husbands death made her heart boil with anger and bitterness. Not necessarily against the Anglo-Saxons, but against Haakon *"Child of Athelstan"*– the man who had ousted them from their throne in the first place. She incited her sons to share her hatred. Together they schemed against Haakon, plotting his downfall day and night.

The Good Reign

Meanwhile by the endless coasts of the North Way, Haakon proved to be a completely different kind of leader than Eric Bloodaxe. Haakon had his values elsewhere, rooted in a different cradle. He commenced his reign by acting upon these values with verve and determination. Knowing that the people lacked a sense of national sentiment, he wanted to lay the foundations of a unified nation and people. The first step was to restore the commoners' trust in the state. He fulfilled his promises of less taxation and abolished several infamous tax laws and regulations, giving the landowners more independent freedom. He spent much time inspecting his subordinates to make sure they treated the civilians with respect and dignity.

The second step was to improve the royal guard, or *hird* – a band of men woven together by a sacred oath to protect and fight for the king and serve his will till death. Any free man could enlist, but had to undergo harsh training and give up many worldly pleasures. The hird was a fellowship of blood brothers; of "*hirdmenn*" (huscarls)[8]. Not only did they swear to defend and avenge the King, but also each other. In return for their unyielding loyalty they were given gifts, prestige, and seats by the king's dinner table. They would feast and drink with him, sharing deep conversations and silly jokes. The King was expected to be the finest warrior of them all and lead them in battle by being the first to engage with the enemy. The grand goal of any *hirdmann*, or *huscarl*, was to die valiantly on the battlefield for their King, country, and faith.

The expansion and organisation of the royal hird was a significant achievement. It became the key instrument for enforcement of the king's will. In other words, it was the main muscle of the state. It also granted Haakon a group of men he could always rely on. Many of the hirdmen became close friends of the King and could often function as advisors, spokespersons or even administrators. In times of war, they were invincible, elite fighters.

By this time, there was no national capital. The king resided on the many royal strongholds that before the unification had functioned as royal seats for the many petty kings and chiefs. King Haakon would travel from household to household with his hirdmen to administer, feast and monitor his subjects. This meant that the king was almost constantly on the move. To accelerate the travelling it was therefore necessary to build a better network of roads. Haakon built the first official web of roads across the country and organized a naval highway between Sogn and Trondelag. The new infrastructure had amazing results. It boosted domestic trade and accelerated commercial activity along the North Way.

---

[8] Hirdmenn were *free men* who protected their leader. In Old Norse, they were called *huskarlar*. *Hus* means house, and *kar* signifies a strong companion. This title has later been anglicised as "*huscarls*" or "*housecarls*".

But what is prosperity without proper defences to protect it with? Haakon knew he had to bolster the defences of the North Way. He organized the entire country into counties and ordered that each county was always to have a predetermined number of trained warriors and longboats ready for mobilization. The concept of national conscription was therefore introduced. Norwegians called this new system *"leidgang,"* and it was used throughout the middle ages. Haakon then completed the erection of hundreds of warning beacons, allowing quick signalling in case of approaching enemies. This enhanced the Kingdom's military preparedness.

What remained was for Haakon to show some military muscle. He attacked Danish settlements in Sweden and raided sporadically in Denmark. Having seized Viken from Danes, he instated his nephew, petty king Tryggve, as ruler of Viken. Such short campaigns helped reduce Danish influence over the area, but it also sent a clear message to the North Way's neighbours not to seek hostilities. Also important was to satisfy the more bloodthirsty Vikings in his realm, test his armed forces and prove to his countrymen that he was a strong, military King. The latter was particularly important as it was expected of any King to be skilled and courageous in combat.

The sagas describe Haakon the Good as a man who sincerely valued the legislature. This is evident in his legal reforms. When Haakon became King, things – assemblies of all free men - were the source of justice. There were plenty of local things, but the larger ones, like the *Gulathing* and *Frostathing,* were the most significant. They called on all free men from numerous counties to come together. However, ever since the unification, these larger things had a range of impracticalities. They could be loud and unorganised, making it hard for the King to voice his will. Farmers would often have their work disrupted as it was their duty to participate in the assemblies. The larger things were simply not effective, so Haakon re-organised them. He essentially turned them into *representative* institutions: each county had one representative that would participate in the assembly. The rest would continue working on their crops or crafts undisrupted. This made everything much easier. It reduced the number of participants in the things, therefore making them easier to manage, scrutinise and oversee. It also enabled Haakon to keep a closer eye with the local chieftains.

Haakon needed this proximity to representatives and chiefs. He wished to introduce new laws, but this had to be carefully conducted in accordance with the farmers. In the Norse world, *only* the farmers had the right to adopt laws, so the King had to negotiate with them in order to have a law implemented. Each thing had a law-code, with the largest one being *the Code of the Gulathing* and *the Code of the Frostathing.* Though it remains unknown to us exactly what laws Haakon introduced, historians Jon Sigurdsson and Synnnøve Hellerud concluded that three of the likeliest laws that came from Haakon were the death penalty for theft, outlawing domestic murderers, and using financial fines as a form of punishment.

Due to his actions, the Kingdom of North Way thrived. An unusual peace fell upon the land, bringing security for merchants, farmers and anglers. Farmers produced more grain than ever before as the earth was abnormally fertile for many years. Legal systems functioned efficiently, and order reigned. The sagas describe Haakon as a remarkably friendly, merry, collected, wise and just man. He had a special talent in uniting people, which was a crucial skill considering the vast amount of quarrelsome Norwegians within the realm. His organisational skills were exceptional, and he was an effective negotiator and speaker. People enjoyed his company. He became known as "Haakon the Good" for his personality as well as his good leadership.

Spark of Christendom

By the time Haakon became King, it is very likely that there were already many Christians on the Western coast. The North Way had attracted many Christian merchants and missionaries to Norwegian shores, influencing the native Norwegians to adopt the Christian faith. However, most of the Norwegian population still belonged to the Norse religion. Haakon, who was a devout Christian, wanted to introduce Christianity to Norway and, hopefully, convert the population. Personally, he always celebrated Christmas and Easter, fasted, and abstained from work on Sundays. In the first years of his reign he would share his Christianity only with his close friends, encouraging them to adopt the faith and abandon their Norsedom. Many eventually did convert.

After Haakon had imported priests from England, he made his intensions clear: he wanted to bring Christianity to *all* of the Norwegians. Numerous churches were built along the coast (like the church on *Edøy* in Maere), and many Norwegians were receptive of the Christian religion. However, when Haakon had churches built more in-land, he was met with fierce resistance. Churches were attacked and burnt down; priests were slaughtered, and Christians were persecuted. Ultimately, the mighty Earl Sigurd of Lade, Haakon's friend, advisor, and ally, protested this new religion together with his Tronds. They had burned a series of churches on Edøy and around Trondelag and demanded that the King explained himself.

Haakon rushed to Trondelag to take care of the issue, but when he arrived, the Tronds were strikingly hostile. Though Earl Sigurd personally sympathised with Haakon, the other Tronds reeked of intolerance. They demanded that Haakon publicly honoured the pagan gods by participating in the Norse religious feast of *blot* (see Terms) [9]. If the King refused, the Tronds vowed they would start an open rebellion and, presumably, continue persecuting Christians. For Haakon, this was a disaster. If he lost the support of the Tronds, he could very easily lose his Kingdom. The Tronds were the core of the army. A Trondic rebellion could throw the country into a destructive civil war and ruin everything he had built.

Earl Sigurd took Haakon to the side to try to persuade him to give in. He would *"gain nothing"* by refusing. Ultimately, Haakon gave in and agreed to join the blot. Through the Norse rituals and feast he was stubborn and partook in few activities. He also made the sign of the cross over every food or drink he consumed, which irritated the others. Earl Sigurd, though he was Norse, defended the King and assured them that he was making the sign of Thor's hammer, not the cross. When it was all over, Haakon was full of fury. He rode back South but swore to raise an army and avenge this humiliation. Earl Sigurd advised him not to, but Haakon would not listen. He began mobilizing an army to attack the Tronds, but he would soon find out that the time was not ripe – the beacons were lit, an external enemy was approaching.

The Battles of Bloodhill and Rastarkalf

Under the influence of their mother, the sons of Eric Bloodaxe were determined to avenge the fate of their father and reclaim the throne. They consulted King Harald Bluetooth of Denmark[10]. Bluetooth also wished to invade, as he was concerned over Haakon's recent aggressiveness, which posed a huge challenge to Danish hegemony in the North. Under King Haakon, the Kingdom of the North Way had begun to take form as an organised medieval country, characterised *not only* by its famous trade coast, but by a distinctive *Norwegian* people with claims to their own country and borders. The *North Way confederation* of Harald Fairhair was turning into the *Norway Kingdom*.

Bluetooth decided to take advantage of the feud and funded the Ericssons (*Eriksønnene;* also called *Sons of Gunnhild*), but under one condition: Norway would be a vassal state of Denmark. The Ericssons accepted. They built a respectable armada of

---

[9] "Blooding" or *bloting* was an old Norse feast that involved religious sacrifices of animals and humans. The blood was spread out in the hall and the flesh was boiled. The participants would then eat and drink to honour the gods.

[10] There is reason to believe Harald Bluetooth was their uncle, but this is not confirmed by historians.

Danish and other Norse warriors, and appointed Guttorm Ericsson, possibly the eldest of the Ericssons, to take command. In 954, they set sail for Norway.

The first attack was in Viken, which was left virtually undefended. After restoring Danish power there, they continued to Agder. Meanwhile, King Haakon had mustered an army and took up position near Avaldsnes, inviting his foe to battle. Guttorm accepted. The two Norse armies clashed violently together, and the battle soon turned into a massacre. After hours of exhausting fighting, the banner of Guttorm was struck down and Guttorm himself severely wounded. With no central command, the invading army lost cohesion and disintegrated. Haakon mercilessly pursued the routing enemy and took no prisoners. Guttorm survived but died of his wounds soon after.

Although this was a crushing victory for Haakon, it was a pyrrhic one. He had lost countless fit and able men that day. So horrific was the sight of the field that it became known as the *Bloodhill* (*Blodheia;* the place still carries this name).

However, Haakon soon learned that he had not yet defeated the invaders: the main Danish fleet, under the command of Gamle Ericsson, had simultaneously attacked Maere further North. The beacons had failed to be lit, causing Haakon to be completely unprepared. As he marched North, he hastily gathered whatever he could find of able fighters to reinforce his battered army.

In Maere, he was approached by an old farmer by the name of Egil Ullserk. Egil had served as Harald Fairhair's flag bearer in many battles and was a tall and stout old man. He volunteered, and his motivation for doing so echoed Norse culture at its finest: *"At one time the peace had lasted so long* [that] *I was afraid I might die the death of old age, but I would rather fall in battle following my King. And now it may so turn out in the end as I wished it to be."* Rather than dying a peaceful death of old age, Egil sought to die a valiant death on the battlefield to earn his place in Valhalla.

With old veterans, some experienced warriors, and a group of hastily recruited farmers, Haakon took up position by Rastarkalf (*Rastarkalv*) in Maere[11]. But it still looked bleak for the King. He had only 400 men to lead against the full force of the Danish fleet. Some of his conscripts lacked armour or even proper weapons. If Haakon was to win the day, he had to rely on deception and luck.

The field of battle consisted of a wide, flat ground on top of a hill. The Norwegian King deployed his men in a stretched line across the ground to avoid encirclement by his numerically superior enemy. It was a fragile line that could not withstand an attack

---

[11] Close to modern-day Kristiansund and Edøy.

for long – but it was all he had. The enemy had arrived and was seen advancing towards them. Suddenly, the invaders charged head-on, and a terrible combat ensued – but Haakon had a card up his sleeve.

As the invaders were fighting, they spotted ten Norwegian military standards appearing from their rear behind a ridge. Gamle Ericsson was taken completely by surprise and, fearing encirclement, ordered his troops in a disorderly retreat. Haakon now had his enemy on the backfoot and led a vicious charge to hunt them down.

But Gamle soon realized that he had been fooled: the ten military standards were carried by just a handful of Norwegians, led by the old Egil Ullserk. They had used the banners to trick the Danes into thinking they were outnumbered and outflanked. Now furious, Gamle rallied his men and charged at the small detachment of Norwegians. However, Gamle's men were confused, unconfident and tired, and most of them had already fled the field. Gamle took his sword and hurled himself at Egil. The two duelled in the midst of the battle, until Egil inflicted a severe blow to Gamle, who fell down and was carried away by his men. Egil, however, had also been struck down, and lay gasping on the field. In that very moment, Haakon's main army appeared and attacked the invaders from the rear, spreading panic among their ranks. With Gamle unable to lead, the army disintegrated into an all-out rout. They fled downhill towards the beach, hoping to board their ships and sail away – but as they arrived they found most of their ships far out in the sea. Haakon's men had pre-emptively seized the beachhead and pushed the ships away. The invaders were trapped. Haakon showed no mercy against them and, again, took no prisoners. Only a handful of them managed to swim to the ships and escape. Gamle was left for dead on the beach.

As the slaughter died down and evening came, Haakon seized the remaining ships from the Ericssons filled them with the fallen warriors. Among them were Egil, who was given a proper funeral. He was given the fate he wished for: to fall in battle fighting for his King.

The battle of Rastarkalf was one of the most cleverly fought battles in Norse history. Haakon had annihilated the enemy force despite being ill-prepared and vastly outnumbered. He had relied on deceptive tactics and Egil's elite warriors to shock the enemy army into complete disarray and cut them down. Norway was defended, and peace could now ensue.

Last Stand at Fitjar

After the battle of Rastarkalf, it took the remaining Ericssons five years to convince Danish King Bluetooth to organise another invasion. Finally, in 960, they had mustered a new fighting force and set sail for Norway once again. This time they were led by Harald *Greycloak* Ericsson, who, after the fall of Guttorm and Gamle, was now the eldest of the brothers.

The fleet moved in stealth and managed to stay undiscovered by the Norwegians. The sons knew where Haakon was, for they sailed directly to Stord, a coastal island that Haakon was visiting at the time. Again, Haakon was taken by surprise and had to improvise an army at all haste. He scraped together what he could find of farmers and took up a defensive position by Fitjar[12] on Stord island.

Greycloak's army was allegedly six times larger than Haakon's, and there was no escape. Sensing the nervousness amongst his troops, Haakon dramatically tore up his royal chain mail and threw it away to inspire the men, then raised his longsword *Kvernbit* and led a charge directly at the Danish force. Hoping to destroy Danish central command, he led his toughest huscarls against Greycloak and the Danish leaders. Haakon, soon stained by blood and dirt, stood in the midst of the calamities. His shining, gilt helmet made him easily spotted on the battlefield – inspiring his own, but attracting fierce attacks from his enemies. A Danish berserker challenged Haakon for a duel and yelled out for the King to show himself. Rather than sending a warrior to go in his stead, Haakon personally appeared and replied, *"Come on as thou art coming, and thou shalt find the king of the Norwegians!"*

The King outmanoeuvred the berserker and struck him down to the enormous cheer of the Norwegian warriors. The remarkable scene emboldened the men, who now lunged at the invaders with brute force. They killed two of Harald Bluetooth's brothers and many other leaders. Pressured by this overwhelming aggression, the Danish army began to crack. They withdrew and boarded their ships to flee the battlefield. Once more, Haakon stood victorious against all odds – however, this time at a heavy price. The King was mortally wounded. As the enemy was fleeing, an arrow had pierced through his ribs underneath his arm. The King was dying and had only a few days left to live. Knowing this, Haakon asked to be carried to his birthplace, where he would prepare to die.

On his deathbed, he was asked who would inherit the throne. Having no sons, yet wanting the Fairhair dynasty to live on undisrupted, Haakon bitterly proclaimed Harald Greycloak his successor. Despite being his mortal enemy, Greycloack was a baptised Christian, and would hopefully continue the work to inspire Christianity in Norway. To ensure a peaceful transition of power, Haakon commanded his subordinates not to resist Greycloack, but accept him as their King and maintain the peace.

---

[12] Can be pronounced *Fit-yar*

One of Haakon's friends promised to bury him in a Christian cemetery, but Haakon implied that since he had participated in the blot rite, he was not worthy enough for such a burial. *"And if fate should prolong my life, I will, at any rate, leave the country and go to a Christian land, and do penance for what I have done against God; but should I die in heathen land, give me the burial you think fit,"* he reportedly said. Shortly after, King Haakon the Good died.

Moved by the bravery their King displayed at Fitjar, poets composed eloquent poetry in his honour. The *Haakonarmal*, a poem describing Haakon's entry into Valhalla, became one of the more celebrated poems in the Norse world.

"The raiser of the storm of shields, the conqueror of battlefields; Haakon the Brave, the warrior's friend; who scatters gold with liberal hand... When gallant Haakon lost his life, black was the day and dire the strife. Amidst the heaps of foemen slain, Haakon was named the bravest on the plain." [13]

Legacy

"So great was the sorrow over Haakon's death, that he was lamented both by friends and enemies; and they said that never again would Norway see such a king."

– Saga-writer Snorri Sturlason.

Haakon was adored by his men-at-arms. He embodied what it meant to be a true Chief in the Norse world: wise and pragmatic, intolerant of cowardice, bold – almost at the point of recklessness, and quick-witted. His cheerful temperament and kind generosity made his friends adore him, and when he died, mourn him. If the Norse Norwegians ought to have a King, they would want a King like Haakon the Good.

King Haakon the Good was a visionary who played a crucial role in turning the North Way into a somewhat cohesive Kingdom. Without him, the unstable kingdom would probably collapse into civil war and have enormous difficulties in reuniting. Having grown up in Anglo-Saxon England and observed King Athelstan since childhood, he had a good idea of how a King ought to rule and how a Kingdom ought to function. He understood that if a country should prosper, it needed infrastructure, legal

---

[13] Merged extracts from skalds Eyvind Skaldaspiller and Thord Sjarekson, as referenced by Icelandic saga-writer Snorri Sturlason.

codes, efficient bureaucracy, a strong military, freedom of the individual, and a just and trustworthy government. All this, he brought to Norway.

However, Haakon did not rule over a strict, feudal society. His abilities as a *Norse* King were severely restricted. When he tried to spread Christianity, he was confronted by the Tronds and later forced to submit to *their* will. This was a sore humiliation for him that he perhaps never recovered from. This episode would later be at the heart of the arguments between Kings and local authorities, as Thing-assemblies would proudly say that they could do with any King as they did with Haakon. But they had to beware – not all Kings would be as mild as Haakon.

Olaf Tryggvason

960 – 1000

Olaf Tryggvason was a legend – by every sense of the word. In the Norse world, he was a celebrity. From the meadows of Ireland to the endless woods of Russia, people told stories of his exploits and adventures, his athleticism, and romantic affairs. He truly embodied the meaning of "hero" of his time and had Vikings from across the Norse world ready to follow him. His enemies, however, despised and feared him. Olaf would first use this influence to win great fortune – but then risk it all to force Norway onto a new path.

Origins

With Haakon the Good dead, Harald Greycloak assumed power in Norway. Initially, he honoured his agreement with Danish King Bluetooth by ruling as a vassal, but after securing complete control over the country, he began departing from his Danish loyalties. Fearful of usurpers, Greycloak commenced a purge of political rivals. The most significant victim was Earl Sigurd of Lade, Haakon the Good's former friend and ally. Greycloak's men locked him in a cabin and torched it, leaving Sigurd to die in the flames. This assassination made a haunting impression on Sigurd's young son, Haakon Sigurdsson, who would soon inherit his father's prestigious title *Earl of Lade* and seek vengeance.

Next, Greycloak hoped to rid himself of Danish influence. He ordered the death of Bluetooth's two vassals in Viken, Gudrød Bjørnsson and Tryggve Olafsson. The death of the latter, in particular, was a blow to Norwegians. Tryggve was a famed Viking, friend of Haakon the Good, and beloved figure. The death of such an icon intensified the already-growing infamy of Harald Greycloak.

Tryggve's wife, Astrid, and infant son, Olaf, fled to Sweden. Greycloak sent word to the King of Swealand asking him to return Olaf to Norway. As the Swedes came to arrest her, Astrid scuffled them away and escaped. She went aboard a ship headed for the Kievan-Rus Kingdom, where she had a brother, Sigurd, servicing the ruler of Kiev. However, as she crossed the Baltic, she was suddenly intercepted by Estonian Vikings who abducted her and seized Olaf. The pirates then sold little Olaf as a slave, and the boy disappeared.

Six years later, Sigurd, Astrid's brother, toured the Baltic coastline to collect taxes for the Kievan-Rus. One day, he spotted a boy that looked rather Nordic. He interrogated the boy, who introduced himself as Olaf, son of Tryggve and Astrid from the Norwegian Fairhair family line. Sigurd realized he had found his long-lost nephew and bought him from the slave owner. He took the boy to Novgorod.

One day, in the marketplace of Novgorod, the nine-year old Olaf spotted the pirate that had originally kidnapped him. Little Olaf grabbed a hatched, ran towards him and hurled it with all his strength at the pirate's skull, killing him momentarily. He then ran away from the scene to tell his uncle Sigurd all about it. In Novgorod at the time, violence within the city walls was illegal and punishable by death. Sigurd was therefore in trouble, and turned to the Queen of the Kievan Rus. He explained everything about the boy and his origins, and the Queen, who sympathised with the boy, pardoned him. Olaf was then adopted by her husband, Vladimir the Great – the ruler of the Kievan-Rus.

Olaf had learned the harsh reality of the world at a very tender age. It gave him an aggressive fighting spirit and a restless mind. The teenager would wrestle and box with the other boys in the area, eventually becoming very athletic. It was clear that Olaf was destined as a warrior. Vladimir therefore gave him the command over the Rus infantry. The soldiers there respected the young lad. His story invoked their sympathies, and his vengeance in the Novgorod marketplace earned their respect. He became quite popular in the army – but, remained unpopular in court. He frequently argued with the ruler, Vladimir. They had a fallout, and a frustrated Olaf decided to leave Novgorod entirely, seeking opportunities elsewhere.

Meanwhile, in Norway, Harald Greycloak faced grim opposition from the Tronds, led by the charismatic Earl Haakon of Lade (who as a boy had witnessed his father, Earl Sigurd, be burned alive by Greycloaks men). Earl Haakon roused the people against the King. After Greycloak attempted to Christianise the country, Earl Haakon usurped the power and ousted the King from the country. Earl Haakon now held dominion over the entire country, but to avoid hostilities with Denmark, agreed to rule Norway as a vassal to the Danish King. Norway was, for the first time, a country without a King.

Olaf the Viking

Olaf led a gathering of Vikings and plundered along coasts of the Baltic. After many raids, he entered Wendland, a pagan Polish kingdom, and agreed to be recruited as a mercenary commander. The sagas reveal that he fell in love with the King's daughter, Geira, and that the two eventually married. This elevated Olaf to the higher ranks of Wendish power hierarchy.

Little is known of Olaf's exact involvements as a Wendish mercenary. Some sources hold that he participated in the war between Emperor Otto III of the Holy Roman Empire and Harald Bluetooth of Denmark, fighting against Denmark. However, this war took place in 974 and Olaf would have been just 14 years old, which makes the claim dubious. Nonetheless, Olaf certainly served on numerous campaigns in Poland, gaining valuable experience and a network of useful connections.

In 990, his Polish adventure ended, as he abruptly decided to leave Wendland and head to Britannia to plunder. The sagas explain his sudden departure by the death of his wife, Geira, who had suffered from a mortal illness. They claim that Olaf was so grief stricken that he did not want to stay in Wendland any longer. Her death also buried Olaf's link to the Wendish royal family and stripped him of further political opportunities, giving him little reason to stay.

As Olaf sailed through the Kattegat sea en-route to Britain, dramatic changes were taking place in both Denmark and Norway. In 964, Bishop Poppa had converted Harald Bluetooth to Christianity, accelerating Denmark's transition into Christendom. In Norway, Earl Haakon of Lade was formally baptised, but he did it for political reasons, and personally remained a devout worshipper of the old gods. He also grew tired of being a vassal to Bluetooth and began to refuse paying tribute to Denmark. Tensions began to resurface between the two Scandinavian countries.

In 987, Harald Bluetooth died and was succeeded by his son, Sweyn Forkbeard. Sweyn went immediately to action: he rallied the so-called *Joms Vikings* – a martial, pagan people residing somewhere around Wolin island – and ordered them to invade Norway to restore the vassalage. By Hjorungavag (*Hjørungavåg*), the Joms Vikings were suddenly confronted by the entire warfleet of Earl Haakon. This would become a spectacular naval engagement and ended in a triumph for Earl Haakon and the Norwegians. The victory wrenched Norway from Danish vassalage and made Earl Haakon its absolute ruler. He therefore became the most powerful Earl in Norwegian history, earning him the fitting epithet *Haakon the Mighty*.

Meanwhile, in 991, Olaf arrived in Britain. These were the isles where endless wars had been fought between Norsemen and Anglo-Saxons. It was not long ago since the devastating invasion of the Great Heathen Army led by Ivar the Boneless, son of the legendary and mythical figure of Ragnar Lodbrok. This massive Norse incursion had united the Anglo-Saxons, who managed to push the Vikings back under the leadership of Alfred the Great. But peace was elusive. Waves of Norsemen kept returning to plunder or occupy lands to settle. Britain was locked in a constant tug of war.

Olaf immediately launched a brutal raiding rampage across England, gathering loot and gaining more Vikings in his ranks. The plan was to wreck so much carnage that he would coerce King Ethelred of Wessex to pay *Danegeld,* a tax paid to the Viking army in exchange for a ceasefire.

Olaf burned himself through Kent and Essex and proved that he was a force to be reckoned with. When he neared Maldon, the second largest city in Essex, the English Earl Britnorth finally confronted him. The Battle of Maldon became a massive brawl, but the Norse Vikings ultimately breached the Saxon shield-wall and routed the enemy. Anglo-Saxon poets describe these Norwegians as wolves and *"beasts of battle."* Britnorth was found dead on the field, and his corpse was decapitated as a stark warning to anyone who dared challenge Olaf and his men. With the road to the lucrative city of Maldon now wide open, King Ethelred finally yielded to Olaf's strict demands. The Anglo-Saxon King emptied his treasury by paying out 10 000 pounds worth of taxes – the largest *Danegeld* ever collected at the time. Satisfied with the success, Olaf went on to raid in Scotland, Ireland and in Northern France.

This Viking warlord was building quite a legacy. His name was on everybody's lips. Contemporaries especially admired his alleged athleticism. They said he could run on ores while his men were rowing, and that he perfected the skill of fighting with two swords. Rumours held he was also a champion wrestler. Vikings from across Britannia came under his service, and many chieftains rallied behind him. At the height of his Viking career, Olaf commanded 94 warships.

He found a useful ally in Sweyn Forkbeard, King of the Danes, who also campaigned in England. The two men joined forces and assaulted London, robbed the Anglo-Saxons of their horses and penetrated deep into Wessex at great speed, threatening Ethelred's political position head on. Contemporaries expected the Anglo-Saxon King to retaliate, but he once again decided to cash out. This time the sum exceeded Olaf's record: 16 000 pounds.

While the English taxpayers were obviously frustrated, they soon learned that Ethelred was not as weak as he could appear. This danegeld came with terms: Olaf Tryggvason was to convert to Christianity, swear an oath never to return to England and become an ally of Wessex. Olaf, surprisingly, agreed. And it was not a trick. Olaf stayed true to his word. He was ceremonially baptised, with Ethelred as his godfather, and cordially terminated his warpath.

It is unclear what Olaf thought of Christendom at this point, but there are indications that he had found a liking of the religion and sought to become more learned in its ethos. After all, the Viking warlord had sometimes used the banner of the cross in his battles. From his caretakers in Novgorod to the churches of England, he had always been in contact with Christians and roamed Christian lands. It is possible that he never felt any special adherence to the Norse religion, and was more naturally inclined to Christendom. Before he reached Britain, he had visited a seer (see Terms) in the Scilly islands and asked him about his future. The seer told him the following prophecy:

"You will become a renowned king and do celebrated deeds. You will bring many men to faith and baptism, and in-so-doing help both yourself and others. So that you may not doubt my words, listen to this sign: when you come to your ships many of your men will conspire against you, and then a battle will follow in which many of your men will fall, and you will be wounded almost to death, and carried upon a shield to your ship, yet after seven days you shall be well of your wounds and immediately you will let yourself be baptised."

According to the sagas, as soon as Olaf returned to his ships, his men mutinied, leading to a combat that mortally wounded Olaf. He was carried away on a shield by his loyal friends and was healed from his wounds after seven days. Seeing all this, Olaf returned to the seer and asked him how he gained this knowledge. The seer replied that he had his knowledge from the God of the Christians, who *"gave him all he needed to know."*[14] Olaf arranged for a personal baptism shortly after.

It is therefore possible that Olaf was already a baptised Christian when Ethelred demanded his official, ceremonial conversion (perhaps Olaf hid his previous Christian conversion from Ethelred to pretend it was a demand to have him baptised, thereby bargaining up the dangeld?). Whatever the case, Olaf would never again return to England to plunder.

Instead, Olaf returned to marry. He heard news that Princess Gyda of Dublin was seeking a new husband. She held a thing to gather all the notable men in the area and pick out whom she fancied. This was, of course, the recipe for envy among men. When she picked Olaf, a jealous admirer challenged him to *holmgang*, a duel to death. Olaf accepted the challenge, but easily outmanoeuvred his opponent. He refused to kill his opponent and let him go, but only after a hard beating. Olaf then married Princess Gyda. He became a landowner in England and seemed to live at relative peace with his new wife.

---

[14] It is speculated that this hermit *seer* was Saint Lide.

Restoring the Throne

Rumours travelled fast and wide. In Norway, Earl Haakon the Mighty became quite curious over who this Olaf Tryggvason figure was. He sent one of his agents, disguised as a merchant, to search for him. But once the agent met Olaf, he betrayed his Earl. Instead of collecting intelligence, he began telling Olaf everything about the state of the country. He said that Earl Haakon had become tyrannical. He allegedly had an insatiable lust for women and forced his subordinates to give up their daughters to him as his concubines. The people began losing their admiration for him. Farmers and other earls now plotted a rebellion.

Olaf saw an unmissable opportunity presenting itself. He could sweep in and assume power just as Haakon the Good had done and reclaim the lost throne of Norway. With all haste, Olaf gathered his veterans, advisors, clerks, and priests and set sail for Norway. In 995, he arrived on Moster island in Western Norway, but hurried further North to reach Trondelag. Upon his arrival there, the rebellion had already started. Olaf rallied the Trondic rebels for a thing and held a passionate speech where he emphasised Earl Haakon's fall from grace. Haakon was, Olaf explained, an Earl, but had assumed powers like that of a King. This was a clear violation of Norse customs and proved that he had been an illegitimate ruler. His words were received with a massive cheer from the rebels, who agreed to hail Olaf as their rightful King. Olaf had a legacy, a proven track record for leadership and a large personal wealth from which he could generously reward new and loyal friends. Moreover, through his father Tryggve, he belonged to the Fairhair dynasty, making him a true Prince of Norway.

The rebels chased Earl Haakon around the country, and the once-mighty man was fleeing from fief to fief. One night, as Haakon was sleeping in a pig barn, his slave assassinated him and cut off his head. He then rode to Olaf and presented it, hoping to be rewarded, but Olaf was disgusted by the slave's actions and sent him instead for execution. Earl Haakon the Mighty of Lade - the champion of the Battle of Hjørungavåg - was betrayed in a pig barn. It was a tragic fate that befell the Mighty Earl.

Olaf headed South to tour the country and secure popular recognition as King. He was especially well received in Viken, as the people celebrated the return of the long-lost son of Tryggve, their former ruler. However, Olaf now began to add another demand at the assemblies: that all of Norway became Christian.

Campaign of Christianization

Why did Olaf require Norway to become a Christian land? Both modern and medieval sources tend to be suspicious of Olaf's intensions. Though we will never know the extent to Olaf's personal faith or how much he knew of Christianity, we do know that he certainly held Christendom – the Christian civilisation – in high regard. While Norway still struggled to keep united, Christian kingdoms like Frankia prospered at pace. Olaf Tryggvason probably sought to bring Norway into the world of Christendom, to become a part of European development and progress. Implied is therefore that the Christianization of Norway was equally a matter of constitutional and cultural reform as it was religious or spiritual.

The Christianisation started in his home region, Viken. This area had already been exposed to Christianity for decades, first through international traders but then through the Danes. The population was therefore baptised with little resistance. A similar success story occurred in Southern Norway. In the Western lands, mass baptism was only arranged after negotiations with the local chiefs. To be approved as King, Olaf married his sister Astrid to Erling Skjalgsson, the most prominent and influential chief in the West. He also appointed Erling as governor of the Western lands. This made Erling the second most powerful man in Norway, a position he would skilfully exploit in later years.

It was only when Olaf returned to Trondelag that he first met serious opposition. As the sagas relate, he was first confronted during a feast. A few prominent, local chiefs approached him and demanded that he sacrificed to the old Norse gods, *or else*. To their surprise, Olaf completely agreed. He even said that he wished to sacrifice to Odin in a spectacular fashion. So spectacular, in fact, that he wanted the finest sacrifices to be made – humans, but not just slaves, he wanted free men to be sacrificed. Then he went on to nominate the local chiefs standing before him, asking them if they would do the grand honour of being sacrificed for Olaf's grandiose blot ritual. This staggering sarcasm was, of course, a threat. The chiefs soon dismissed the idea of a blot and instead agreed to be baptised.

However, this was just the beginning of Olaf's clash with the Tronds. A few days after, a thing assembly was held in Trondelag where Olaf discussed with the people. He was immediately met with the same hostility as before: the angry Norse farmers demanded of Olaf to participate in the *blot* ritual and honour the old gods. *"This we did with Haakon the Child of Athelstan...and we do not regard you any higher than him!"*

Olaf soon had an army of Tronds against him, led by the influential landowner they called Ironbeard (*Jernskjegge*). After a loud thing assembly, Olaf seemed to buckle under pressure. Again, he seemed to agree to engage in *blot*. Chiefs from across Trondelag now gathered to witness the humiliation of the King – but as soon as Olaf entered the Norse temple, he *suddenly* swung his axe on the grand statue of Thor. His huscarls proceeded to smash the other pagan figures. The King immediately arrested Ironbeard and had him murdered outside the temple. Then, he faced the shocked Norse crowd and gave them a chilling ultimatum: either submit to the King and his faith or face total war. None of the other chiefs dared to speak against him, and they were all baptised instead.

Needless to say, behind the following baptism lay a deep grudge against Olaf. The Tronds had always been very independent and despised all who attempted to rule them. The new Earl of Lade, Eric Haakonsson, the son of Earl Haakon the Mighty (!), was hungry for vengeance. In an attempt to appease the hostile people, Olaf married Ironbeard's daughter, Gudrun, as a token of reconciliation. But when Gudrun tried to slit Olaf's throat during their wedding night, the marriage naturally came to an abrupt end.

The key to the throne of Norway lay in Trondelag, so no matter how difficult relations were, Olaf had to ensure control and stability in the region. He founded a trading city called *Nidaros* and made it the capital of the Kingdom. The town was well situated, and soon became a flourishing city. The King also ordered the construction of a gigantic dragon-ship he called *The Long Serpent* (*Ormen Lange*). It was the most formidable warship in his day, manned by a carefully selected crew. *"None were allowed to crew it unless he had proved himself in great deeds...on that ship existed only giants, not whimps or cowards,"* the sagas tell us.

Olaf ensured the Christianisation of the other Norwegian communities across the North Sea. The Faeroes and Orkneys were Christianised under his reign. The most extraordinary instance took place on Iceland, where all its chiefs met at their *Althing*, heard King Olaf's advocacy, and voted in favour of accepting Christendom. Pagan rituals were still admitted so long as they did not "disturb" the community.

However, in the Far North of Norway, it did not run as smoothly. King Olaf brought an army with him and occasionally used 'sticks and carrots' to baptise unwilling Northmen. For example, chieftain Hárek of Tjotta (*Hårek av Tjøtta*) first mustered an army to resist Olaf, but when he gave way, Olaf luxuriously rewarded him. In 998, Olaf encountered the Norse seer Raud the Strong. According to the sagas, Raud was so blasphemous that he triggered Olaf's hellish rage. Olaf and his huscarls captured Raud and tortured him to death. The King eventually retired from the Far North, but planned to return later on.

Discovery of America

At this time, King Olaf was visited by a certain Leif Ericsson, the son of the famous explorer Eric the Red. Eric the Red had once been banned from Norway over a blood feud, but then banned again from Iceland for the same reason. He was clearly an ill-tempered man – but also very bold. In 982, Eric the Red had gambled on sailing further North-West for Iceland in an attempt to discover a new island, rumoured to exist somewhere in the ocean. To his great luck, Eric discovered a vast island with plenty of good soil and green lands. He called it *Greenland*, as *"people would be attracted to go there if it had a favourable name,"* Eric himself admits according to the sagas. He then led a vast migration of impoverished Icelanders to settle there to start a new life. Soon, scores of other brave souls came to settle.

King Olaf now wanted to learn about Greenland, and asked Leif Ericsson to tell him about it. The two spoke candidly together. When he understood that Leif was a Christian, King Olaf admitted him into his personal hird and tasked him with the mission of preaching Christianity to the Norse settlements of Greenland.

With a cross around his neck, Leif Ericsson left Norway and set sail home to Greenland. However, he was caught by strong winds and a storm, making him accidentally drift much further West. In year 1000, he encountered a strange, new land there. He probed into it and found evidence of human life, and by the beach he found two shipwrecked seafarers. Utterly perplexed, Leif left and navigated to Greenland. He then investigated all the latest rumours by seafarers and found one peculiar sailor who claimed to have seen the same land. Leif realised he might have made a greater discovery than his father!

With a new crew and ship, Leif set out again. They first found a desolate wasteland they called *Helluland* (perhaps the Baffin islands), then a land full of woods, which they called *Markland*. Finally, they settled further South in a fruitful area filled with grape-trees, calling it *Vinland* (meaning Wine-land). Here, the earth was fertile, enriched by plenty of salmon-populated rivers and grapes. They set up a Norse colony here, while Leif returned to Greenland to report the discovery. He became known as *Leif the Lucky.*

However, Leif the Lucky had made a promise to King Olaf Tryggvason to promote Christianity on Greenland, and he kept true to his word. Instead of exploring Vinland further, he stayed in Greenland for the rest of his life, preaching Christianity to the best of his abilities. Eric the Red was sceptical and stuck with his old Norse gods, but Leif's mother, Thodhild, vigorously accepted Christianity. She became a staunch advocate of the religion and even built churches across the colony. Monasteries also sprung up. Thus, with Olaf's decision to send the lucky Leif Ericsson as the missionary of Greenland – he led Leif into his fortunate fate of becoming the first Western European to discover America, five centuries before the days of Christopher Columbus.

The Norse colony of Vinland later engaged with the Native Americans, whom the Vikings called *skrælinger*. The natives were hostile, and many battles were fought between them. Leif's brothers and sisters would continue to lead expeditions into America and fight against the natives, but this often lead to an unknown fate. The colony survived for some time, but the distance and dangers of the journey to reach it, coupled with the hostility of the natives and other challenges, made the Vikings eventually abandon it.

Battle of Svolder

Meanwhile, back in Scandinavia, it did not look as good. The tide was generally turning against Olaf Tryggvason. He failed to secure marriage with Queen Sigrid the Haughty of Swealand, who despised him after his attempts to convert her. Instead, Olaf eloped with Princess Tyra, King Sweyn Forkbeard's sister. This angered the Danish King and gave him an excuse to raise hostilities against Olaf. Sweyn already had grievances with him. Ever since Olaf's departure from England, Sweyn had struggled to uphold the front against Ethelred, which agonised him. Olaf's restoration of the Norwegian Kingdom also undermined Danish claims on Viken. This concern was shared with King Olof Skottkonung of Swealand.

Earl Eric of Lade took the initiative to form a coalition against Olaf. By 999, therefore, the King of Swealand, King of Denmark, and Earl of Lade had all united to force the downfall of Olaf Tryggvason.

In year 1000, Olaf was sailing cross the Baltic on his way back to Norway after settling a dispute in Wendland. He had one of his most skilled subordinates, Einar Thambaskelfer of Trondelag[15], by his side. Abruptly, a huge enemy fleet appeared in an ambush. Olaf allegedly had only 11 ships, while the enemy fleet had 80. When some of his men suggested he should flee, Olaf allegedly replied: *"Strike the sails! Never shall men of mine think of flight. I never fled from battle. Let God dispose of my life, but flight I shall never take."* He organised his fleet in a tight formation, roping the ships together so they likened a floating fortress (same strategy as Harald Fairhair on Hafrsfjord). The enemy warfleet then ripped into their formation, and terrible combat ensued. Olaf's Long Serpent was the tallest and largest ship, enabling his men to shoot arrows and javelins from a highpoint. The missiles rained down on the coalition and caused them staggering losses. Soon, both the Danish and Swedish fleets disengaged and resorted to skirmishing – leaving the battle between King Olaf and Earl Eric of Lade.

---

[15] Thambaskelfer or *Tambarskjelve* is an Old Norse cognomen of unclear meaning and may mean either *the Longbowman* or *Pot-bellied*.

At this stage, the coalition finally had a breakthrough as they began clearing several of Olaf's ships. They then poured all their men on the Long Serpent, filling it with as many men they could, and driving Olaf's men to their doom, one by one. Olaf and his companion Einar fought desperately – but they now understood that all hope was lost. *"What broke with such noise?"* Olaf asked as he heard Einar's bow crack. Einar cried back: *"Norway, king, from thy hands!"* After an exhausting struggle, Olaf was cornered at the edge of the ship and it was clear that his end had come. Suddenly however, he jumped aboard and slipped into the sea. Many of his companions did the same, although Einar stayed and surrendered. The coalition forces shot arrows and spears at their swimming enemies, and many fell either from drowning or being hit. They then searched for King Olaf but couldn't find him. He mysteriously disappeared and would never return.

After the defining battle of Svolder, Norway was divided between the members of the coalition. The Danes re-assumed power over Viken; the Swedes gained parts of Norway's eastern territories and Earl Eric of Lade ruled the rest of the country together with his brother, Svenn. Meanwhile, Erling Skjalgsson enhanced his influence as the effective ruler of Western Norway. The Kingdom of Norway went in the shadows.

What happened to Olaf? No one has the answer. Some said that he drowned and sank to the bottom of the Baltic. Others claim he managed to rid himself of his chainmail and swim ashore. The *Flatøybok,* a medieval compilation of sagas, claim he renounced all earthly ambitions and became a monk near Jerusalem. Others would later do pilgrimage to the Holy Land and claim to have seen or spoken to him. But some dispute this, claiming that Olaf could not have survived the many wounds he suffered at the battle of Svolder. A veteran from the battle sung: *"Does Olaf live, or is he dead? Has he the hungry ravens fed? I scarcely know what I should say, for many tell the tale each way. This I can say, nor fear to lie, that he was wounded grievously. So wounded in this bloody strife, he scarce could come away with life."* (Halfred Ottarson, navigator aboard *Long Serpent*)

Whichever the case, Olaf disappeared from history.

Legacy

"...and thus he made such a fortunate advance in his undertakings, for some obeyed his will out of the friendliest zeal, and others out of dread,"

– Snorri relates.

Olaf Tryggvason is considered to be one of the most important kings of Norway. He had an unbelievably adventurous life story, starting with the chains of slavery and ending as the King of Norway. His memory was especially revived in the 1800's, when he served as a source of patriotic inspiration.

Yet, his reign lasted only for five years. How could these few years mean so much to Norwegians? The obvious central element of his reign was his Christianisation. No one can deny that he was eager to bring the Christian civilisation to Norway. The consequences of his religious policies were that large parts of the country formally converted, accelerating a process of national development, cultural revolution, and social reconstruction.

Some have, of course, trouble approving his methods of conversion, as he sometimes turned to violence. This is very understandable, as it was clearly at odds with the virtues of Christianity. To get the full assessment, let us remind ourselves of the context.

Firstly, along with Olaf's calls for mass-baptism were his calls of public recognition as King of Norway. This meant that his campaigns were both politically *and* culturally motivated. He did not demand monarchical recognition and baptism in separate, but jointly. It meant that those who denied were viewed both as enemies of the King and of Christendom. But what about those occasions when Olaf *only* called for mass-baptism and still coerced? Here we should remind ourselves of the context. This was a harsh Viking culture that respected display of force. We have only to look at *Olaf Tryggvason's Saga,* written by Snorre Sturlason in the 13th century. Though the saga is written to *honour* Olaf, it includes ghastly stories of torture of the King's enemies and sometimes even *exaggerated* his brutality. This implies that violence, when viewed as a "righteous" necessity (based on contemporary standards) was not a pejorative concept. As a King, Olaf was *expected* to be harsh to his enemies and mild to his friends.

Besides, violence was the language Olaf spoke. His childhood was filled with traumatic experiences of being abducted and sold away by strange, hostile adults. He murdered at the age of nine, and lived a life of Viking raids and wars before his conversion to Christianity. Olaf grew up a warrior, lived like a warrior and therefore ruled as a warrior. This was simply a very different age and time. His methods were certainly no different from that of other Norse chieftains at the time.

The myths and legends of Olaf continued to reach folks across the medieval world. To the Icelanders, who advocated for him to be venerated a Saint, he was a pious and serene Holy Man. To the Anglo-Saxons, a terrorising warlord responsible for many atrocities. German churchmen claimed he was a heretic who had a distorted understanding of Christianity. Contemporary Norwegians hailed him as a strong King and warhero. Modern historians usually like to portray him as a tyrant. All these viewpoints have made him one of the most exciting, debated, and intriguing characters in Norwegian history.

But the work he had begun was not fully completed. The ideas he had sat in motion where permeating the masses, but not yet materialised. Maybe Norway could unite properly, and stay independent? Maybe it could join the richness of the European civilisation, and, like Frankia, the Holy Roman Empire, the Byzantine Empire and England, become a strong, cohesive nation? All the country needed was another Olaf to finish the work the former had begun.

Saint Olaf

ca. 995 – 1030

"It is the honour of a King to defeat his enemies, but an honourable death to fall fighting with his men."

– Anonymous, quoted from *Olaf Haraldsson's Saga*.

The Eternal King of Norway, as he is known, is not only the patron saint of Norway, but one of the most important and legendary characters in Norwegian history. His reign was filled with important reforms, helping the constitutional fabric of the country solidify and tying the country together to form a tighter, national union. Similar to Tryggvason, he held the characteristics Norwegians at the time admired, making him a hero in the memory of the people. This was the man who completed the Christianisation of Norway and changed Norway forever. Through his dramatic life, and ultimate sacrifice, he became the national symbol of patriotism.

The Stout Viking

Olaf Haraldsson's life began in Viken. His father was Harald Grenske, a petty king of Vestfold and Agder who had served as a Danish vassal. Harald Grenske was also the great-grandson of Harald Fairhair. However, he died shortly after Olaf's birth, but his wife Aasta (Åsta) quickly re-married. The stepfather was Sigurd the Sower (Sigurd Syr) the petty king of Ringerike and, also, great-grandson of Harald Fairhair. Sigurd was known to be a wise and prudent man and exerted significant influence in Eastern Norway. He had converted to Christianity in 998 under Olaf Tryggvason's reign.

At age twelve, Olaf was sent on a Viking expedition under the care of the Viking warrior Rane. The expedition took the young boy across the Baltic Sea, exposing him to harsh raids at a very tender age. He did not return to his parents but stayed *in Viking* instead. As years passed, Olaf grew to become a relentless warrior. He had flaming red hair and was very large in stature, earning him the epithet *Olaf the Stout*. He led many daring Viking expeditions across Europe, even reaching Spain. His band of Vikings would raze villages to the ground, loot, rape, and plunder.

In 1010, Olaf the Stout joined Thorkell the Tall, a Danish warlord, and his armada of Vikings. Together, they assaulted East Anglia, the Midlands and Wessex. They attacked London, where Olaf famously sabotaged one of the bridges over the river Thames, making it collapse and sending scores of Anglo-Saxon defenders to the depths of the river. Yet the most unforgiving attack was on Canterbury. The city was wrecked, civilians butchered, and the Vikings even captured the Archbishop, who was later killed by a group of drunken warriors. Despaired and desperate, King Ethelred paid a staggering sum of 48 000 pounds of silver for a ceasefire. Olaf received a good share of the money.

As part of this agreement, Olaf and Thorkell agreed to switch sides and fight for Ethelred. Olaf's first deployment was in Northern France, where he was lent out to Ethelred's ally, Duke Richard II of Normandy[16]. Olaf and the Duke formed an unexpected friendship. Since they spoke the same language, they frequently conversed and discussed various topics. It is probable that they discussed Charlemagne and his successful Empire-building, the principles of European feudalism and the Christian civilisation. They visited the Cathedral in Rouen together, where they discussed Christian theology with the Archbishop. The sagas further relate to a story when Olaf had a dream of a *"great and important man...*[of] *terrible appearance"* who spoke to him saying: *"Return back to thy udal* (ancestral land)*, for thou shalt be King over Norway for ever."* In 1013, Olaf Haraldsson was baptised.

When Olaf returned to England, he received the disruptive news that Sweyn Forkbeard had died. This encouraged Ethelred to open a massive counteroffensive against the Danes. Olaf played a significant role. Through 1014, he helped the Anglo-Saxons drive the Danes back into the sea. Peace finally resettled, but not for long. In 1015, the new King of Denmark, Cnut, led a new invasion of both Danes and Norwegian warriors. They swept across Britannia at an unstoppable pace and were determined to finally conquer all of England for good.

The Reconquest

Earl Eric of Lade, who had now governed much of Norway for fifteen years, was one of King Cnut's chief subordinates. Cnut favoured Eric so much that he gave him the

---

[16] Remember, this was the same Normandy that was founded by Rollo, or Rolf the Walker, following Harald Fairhair's conquest of Norway. Even at Olaf's time, some Normans probably still spoke the Norse tongue.

Earldom of Northumbria – the heart of Danelaw. Earl Eric thus spent the rest of his career sitting in York. He entrusted his brother Earl Svein[17] and his son, Earl Haakon the Younger of Lade, to rule Norway in his place. However, this was about to change.

In the same year of Cnut's invasion, Olaf the Stout had set out to return to Norway. He had about 300 warriors with him, a handful of priests and friends. His goal was ambitious to say the least: he aimed to wrench Norway from foreign control, complete the Christianisation of Olaf Tryggvason and implement a new constitutional and political regime, like those of the rest of Europe. With King Cnut and Earl Eric both hotly tied up with wars in England, the timing was ideal. He had an abundance of riches from his raiding in Europe and a solid gang of experienced warriors.

Olaf first landed in Western Norway. His first landing gave rise to a charming anecdote. As he walked his first steps back on Norwegian soil, he stumbled, and said *"I fell!"* One of his huscarls replied: *"No, you gained your foothold in the land."*

They then proceeded in stealth to track down and ambush the young Earl Haakon of Lade, hoping to catch him before he could muster his men. Having located him, they lured the young Earl into a trap and, once captured, gave him an ultimatum: either face death or publicly announce Olaf as the new King of Norway. The petrified Haakon chose the latter and partook in a series of public events where he, in front of witnesses and crowds, hailed Olaf as King.

With these initial strategic moves, he had a strong foundation to build on. But before any further politics could resume, Olaf sought counsel with his stepfather and mother. He went home.

The sagas describe the remarkable scene of how Olaf arrived ahead of hundreds of strong, stout, and scarred veterans. It must have been quite a sight for his parents, who had not seen him since he was twelve years old. In the ensuing conversation, Olaf made clear his ambitious plans: *"To say the truth, I intend rather to seek my patrimony with battle-axe and sword, and that with the help of all my friends and relations, and of those who in this business will take my side. And in this matter I will so lay hand to the work that one of two things shall happen, -- either I shall lay all this kingdom under my rule which they got into their hands by the slaughter of my kinsman Olaf Tryggvason, or I shall fall here upon my inheritance in the land of my fathers."* Olaf then asked Sigurd to use his extensive political network to arrange for thing-assemblies in the Uplands and other

---

[17] This name is spelled in the Norwegian manner as opposed to the anglicised *Sweyn*. This is to avoid confusion between the characters.

regions. "*It is no small affair, King Olaf, thou hast in thy mind,*" Sigurd remarked. He warned Olaf of the dangers he was putting himself in but agreed to help him in all affairs and in every way. He tied his fate to Olaf's.

Thing assemblies were arranged in various parts of the country, where Olaf spoke and convinced them of his legitimacy and Norway's right to self-rule. He secured the Uplands and Viken, two important regions for manpower. His next challenge was then to tackle Earl Svein, who was now backed by Earl Einar Thambaskelfer (the former companion of Olaf Tryggvason and survivor of the battle of Svolder) in Trondelag. These men would not give up without a fight.

Svein mobilised a Trondic army and sailed into Viken, aiming to strike at the heart of Olaf's dominion. A major naval engagement ensued near Nesjar, where Olaf promised his warriors generous rewards in gold if they proved their worth. The men threw themselves into a ferocious attack but were repulsed by the stubborn Tronds. The tide of the battle swung back and forth for hours on end, until the casualties proved too many for Svein, who admitted defeat. It was a hard-fought victory for Olaf, but a victory nonetheless.

The road into Nidaros, the capital of Norway, lay wide open. Olaf sailed in and occupied the city, taking Einar Thambaskelfer captive, but choosing to pardon him. He then received news that Earl Svein had died in Sweden of unknown causes. All of Norway was within his grasp – but he could not celebrate too early. There was just one more knot to solve: Erling Skjalgsson, the governor of the Western lands. Nothing was promised until he could find a way to subdue him.

There was no sweet-talking in the meeting between Olaf and Erling. Olaf presented the details of his new regime, loud and clear, and Erling totally rejected the terms. The new regime, if implemented, would effectively rob Erling of the power he had accumulated so far. The two men almost brawled, but were appeased by their advisors. However, Olaf had the upper hand. After all, he held the popular vote in virtually all of the things in the country. Erling, seeing that it was impossible to change Olaf's mind, bitterly accepted the terms. With Erling temporarily subdued, Norway belonged to Olaf the Stout.

Regime Change

Olaf Haraldsson's reign differed from previous ones. Deeply influenced by Charlemagne and the feudal Kingdoms of Europe, Olaf sought to implement the *European model* to Norway. Among other things, this meant centralising power to himself as King.

Of course, this was very controversial as it trespassed the conventional boundaries of a Norse monarch. As we know from our introduction, Norway was traditionally governed by localities: thing assemblies, independent farmers, and petty kings. Olaf's road to centralisation was riddled with opposition.

Before he went ahead with his constitutional reforms however, he sought to complete the Christianisation of his predecessor. While the rest of the country had more or less remained Christianised, the Uplands and Trondelag had relapsed into Norse paganism. Already, the Uplands were in rebellion. Olaf ordered his agents to capture the rebel leaders while asleep, torture them and release them back into society as a chilling warning to anyone who had similar ideas. The King struck with an iron fist.

In 1020, he headed to the Far North to establish power there. Each thing assembly agreed to vote him in as King, and when the prominent chiefs Hárek of Tjotta and Tore the Hound (*Thorir Hund*) swore their allegiances, Olaf had finally brought the Far North into the realm. For the first time, virtually *all* of the Norwegian lands were under the authority of a single King[18] - Olaf the Stout held more power than any other Norwegian King before him.

Trondelag, as usual, was harder to subjugate. Blot rites continued and the Tronds kept Olaf at bay through deception and diplomacy. One day, Olaf had enough. He arrested their leader Olve, confiscated their lands and wealth, exiled numerous persons and forcefully implemented his own network of governors. The most notable of these new governors was Earl Kalf Arnesson, one of Olaf's most trusted companions, who was bestowed the governorship of all Trondelag. Similar reprisals were taken against other rebellious kings in Hedemark and Gudbrandsdalen.

What happened in Trondelag is a classic example of how King Olaf centralised the country. He pushed his own direct authority nationwide, sacking those who protested. The only person he truly struggled to fully humble was Erling Skjalgsson, who continuously tried to restore his former powers. Olaf and Erling frequently feuded, as both stubbornly refused to back down. It was just a matter of time before they would clash violently.

Theoretically, the Christianisation of Norway was complete at this point. However, folks knew little about the actual Christian faith. In some cases, they worshipped *Kvite-Krist* (Norse name for Jesus Christ, directly translated *White-Christ*) alongside their Norse gods. It was therefore necessary to establish a national church institution that

---

[18] As it is disputed to what extent Harald Fairhair really controlled the Far North, or Halogaland.

could hold mass every Sunday and preach the pillars of Christianity. As a monarch, Olaf additionally wanted to construct a new, national legal code, based on Christian virtues.

Around 1022/23, Olaf held a large thing assembly at the island of Moster – a symbolic place as it was here Olaf Tryggvason founded his first church. Attending the thing were priests, theologians, state servants and Bishop Grimkell, Olaf's religious mentor. They dictated a new law code they named *"Kristenretten"* (the Christian Law). It prohibited polygamy, adultery, wizardry, abortion[19], blood-vengeance, pagan rituals and specified what was required of every Christian. It also established the rules and procedures of the Church. In accordance with this, Olaf systematised the building of churches by laying out how they should be spread across Norway to preach Christianity. The thing assembly at Moster unified Church and State, thus formally founding the Church of Norway and officially making Norway a Christian Kingdom.

Along with the Christian Law came a formal implementation of feudalism – an enhancement of Olaf's grip on power at the expense of Earls, chiefs and petty kings. In accordance with this, Olaf confiscated large farms and properties and allocated their ownership to the State or the Church. Old chieftains and local rulers were replaced with *lendmenn* – local governors appointed by the King - and *årmenn* – state servants who monitored and scrutinised the governors.

This shook the entire country, as it rattled the heart of Norse customs and traditions. The most controversial was his property confiscations. It gave him many bitter enemies.

However, many argue that such controversial actions were necessary to modernise the Kingdom. If the King's and Church's position were to be elevated, they *needed* land. Land ownership was the currency of the day. Also, though it outraged the Norwegians, it was nothing but the standard, feudal social structure of medieval Europe. Olaf Haraldsson emulated the European monarchs, making Norway a Kingdom of an absolute monarch, supported by a clergy and an aristocracy. This may have made sense to other international Norwegians – but for the average farmer, it was an outrage. This modern Europeanism certainly was not so charming to them.

Geopolitical Wars

The power dynamics of Scandinavia were changing. Thanks to Olaf's consolidation, the Kingdom of Norway consisted of vast swaths of territory in mainland Scandinavia, in

---

[19] In Norse custom, abortion was carried out by deserting unwanted infants in the woods to be eaten alive by wolves or starve to death.

addition to the Orkneys. To the East, Swealand and Geatland united under King Olof Skottkonung, establishing the unified Kingdom of Sweden for the first time. To the South, King Cnut of Denmark had made enormous progress in England, carving out a *North Sea Empire.*

Olaf's relations with Sweden were already quite poor. The Norwegian King had reclaimed authority over the Uplands, which had briefly been under Swedish taxation after the fall of Olaf Tryggvason. Frequent skirmishes took place on the border. To make matters worse, both Kings were ill tempered and despised each other. In 1019, the skirmishes nearly escalated into total war, but it was hindered by state servants and lawyers on both sides, who managed to appease the Swedish King and facilitate a peace treaty. As part of the treaty, Olaf married Princess Astrid of Sweden, who borne him a daughter. Shortly after, Olof died, leaving the power to his milder son, Anund Jacob.

In 1024, messengers from King Cnut of Denmark arrived at Olaf's court in Tonsberg (*Tønsberg*). Their proposition was for Olaf to become a vassal of King Cnut, in exchange for an abundance of wealth and privileges. Olaf flatly rejected their proposal, countering them by emphasising Norway's rightful freedom from foreign rule. He must have been aware of the consequences of such a reply. At the time, Cnut was the most powerful man in the North. He had succeeded in conquering *all* of England, making his North Sea empire span from Wales to Scania. It earned him the prestigious epithet *Cnut the Great.* King Olaf hastened from thing to thing, recruiting free men to join his army against the inevitable grand war against Cnut's empire.

Gathering forces proved to be harder than expected. His controversial policies had made him unpopular. People flocked to Erling Skjalgsson, who obviously exploited the situation and defected to Cnut. Other key players also defected, like Tore the Hound from the North. Agents of Cnut were everywhere, offering fat bribes to Norwegians. The poet Sigvat the Skald sung: *"the King's enemies are walking about with open purses; men offer their heavy metal for the priceless head of the King."* This poet, Sigvat, had personally done the reverse. Learning of Cnut's intensions, he abandoned the Danish court and came to Olaf. He regarded it as his duty to remain loyal to *"my own Norway King,"* and was deeply repulsed by those who betrayed Norway's independence for gold.

> *"Our men are few, our ships are small, while England's king is strong in all; But yet our King is not afraid - O! never be such King betrayed!"*
>
> – Sigvat the Skald sung.

Olaf was lacking in manpower compared to Cnut, but he was luckily aided by Swedish King Anund Jacob. Besides being Olaf's brother-in-law through Olaf's Swedish wife, Anund felt threatened by Cnut's ever-expanding empire. United by a common enemy, the two regents joined forces and attacked in 1026. They invaded Denmark and sacked its towns and fiefs for resources.

Cnut mustered his warfleet and confronted them at Helgeå – *Holy River*. Cnut's fleet was enormous – allegedly tallying 600 ships – while the combined Norwegian and Swedish fleet consisted of around 480. After a gruesome clash and a series of tactical manoeuvres that led to huge casualties on both sides, Olaf and Anund Jacob were forced to pull back. Sagas and chronicles dispute the exact result of the battle – some say coalition victory, others give it to Cnut. Most likely, Cnut technically won the battle, but with staggering casualties that impaired his ability to continue the campaign. In the meantime, the Norwegian-Swedish alliance dissolved, and Olaf returned to Norway (sagas claim Anund Jacob feared having his military strength annihilated through these costly engagements with Cnut, and therefore withdrew).

One year later, in 1028, King Cnut invaded Norway ahead of 70 warships – this time with sweeping success thanks to the cooperation of Norwegian nobles. They toured in the Western lands where Erling accommodated them. Cnut then proceeded to Nidaros where he had himself officially crowned King of Norway. He met Earl Einar Thambaskelfer there, who also defected to Cnut. Meanwhile, Olaf tried desperately to gather all the men he could – but Cnut's gold ran deep in Norwegian pockets.

An opportunity to attack presented itself when Erling Skjalgsson neared Jæren in Viken. Erling had taken the matter in his own hands, and had come to kill Olaf. But this was reckless. Viken was Olaf's home. He knew every river and lake, and lured Erling deeper into a trap at Bokn in Ryfylke. Here, Olaf's men suddenly charged into Erling's fleet and destroyed it.

Erling himself was captured and Olaf's huscarls proudly presented him to their King. It was a tense moment. Finally, Olaf had his age-long rival on his knees. Olaf took his axe, and gently placed it on Erling's chin. But to the surprise of his men, refrained from killing. He pardoned Erling and walked away. Erling was simply too popular to kill. Killing him would cause too much of an uproar in the public. Yet in that very moment, Olaf's huscarl Aslak unnerved and hurled his axe at Erling, killing him instantly. Olaf cried out in fury: *"You fool! Now you hewed Norway off my hands!"*

The King was accurate in his prediction. The execution of Erling Skjalgsson trigger a major uproar across the country, giving rise to an open rebellion against King Olaf. He now found himself deserted on all sides. He had to admit that he had lost the Kingdom. In bitterness and grief, he fled the country together with the last of his companion huscarls and his family. Cnut's conquest of Norway was then complete. He put Earl Haakon the Younger back in as governor of Norway on his behalf.

Return and the Battle of Stiklestad

In 1029, Olaf arrived to Kiev, where its ruler Yaroslav the Wise warmly welcomed him. He told him all that had happened. Yaroslav offered Olaf the throne of Bulgar [20], but Olaf declined. His future lay not in Eastern Europe, for he longed back to his homeland.

However, his prospects of reclaiming kingship over Norway were meagre. He did not have the resources or influence to build an army capable of matching Cnut. Nor did he have any support with the people, who had turned against him. Venturing into Norway again would be suicidal. Olaf was in deep despair. For many nights, he lay awake, haunted by ill thoughts.

But one night he was visited by a *"tall and glorious"* man (as the sagas describe it), dressed in beautiful robes. The overwhelming appearance of him struck Olaf with humility and fear. He trembled and listened carefully. He was strictly told to return to his Kingdom and fight for the throne God had given him. *"Do not let your inferiors frighten you. It is honourable for a King to rise victorious over his foes, but an honourable death to fall by his men in battle!"*

This spectacular, supernatural moment completely rejuvenated Olaf's spirit. Ruled by zealous conviction, he immediately sought an opportunity to return to Norway one last time, just as the glorious man had ordered him. He then heard news from the North: Earl Haakon the Younger, the governor of Norway, had died, leaving Norway temporarily leaderless. Olaf knew this was his God-given sign to return. He rallied his 200 companions and left Kiev immediately, heading North to Norway. Make no mistake – the chances for success were slim, but it did not seem to matter anymore. Olaf had a divine call to act on.

He recruited more troops along the way so that upon his arrival into Norway, his army had reportedly grown to 4000 men. Some were old veterans and friends, others were enthusiastic recruits from Russia and Sweden. There was also a clan of atheists in Sweden who agreed to be ceremonially baptised in order to follow him. Olaf's 15-year-

---

[20] Not to be mistaken as Bulgaria. The throne of Bulgar held power over Bulgarian lands by the Volga river, where the city Bulgar lies today, near Kazan.

old half-brother, Harald Sigurdsson (later known as *Harald Hardrada*) also came along to fight.

As Olaf approached Nidaros however, he was blocked by a massive opposition army, consisting of farmers from Trondelag, Maere and Lade. It allegedly tallied 8000 to 14 000 farmers - a staggering number and the *largest* army ever assembled in Norway up to that point [21]. Leading it were old friends of the King: Horic of Tjotta, Tore the Hound and, to Olaf's disappointment, his former huscarl Kalf Arnesson.

Olaf's subordinates advised him to pillage and burn the farmlands to punish the farmers for their disloyalty, but Olaf refused. He sought reconciliation with his enemies, not more war and destruction. The days of civil strife ought to end, not perpetuate.

On the 29th of July, 1030, the two armies faced each other by Stiklestad, north of Nidaros. The opposition army was flanked by two rivers, while Olaf's army stood on a wide plain. Hoping to sap the enemy's advantage in numbers, Olaf planned to pin the enemy army between the rivers, while sending his Marshal Dag Ringsson and his men to attack the enemy left flank. Dag would push the flank until its collapse, then swing into the centre and finish the enemy. It was an ambitious plan, especially considering Olaf's numerical inferiority. He knew it all depended on the ferociousness of his men.

As the two forces approached each other, Olaf spotted Kalf, he said to him: *"Why are you here, Kalf? In Maere, we parted as friends."* Kalf replied: *"Many things have changed. You parted from us so that it was necessary to seek peace with those who were left behind in the country..."* Finn, one of the Olaf's loyal huscarls, rebuked him: *"This is to be observed of Kalf, that when he speaks fairly, he intends evil."* There was another exchange, until Olaf closed with the words: *"...fate will not give you victory today over me, who raised you to power and dignity when you were a small man."*

Olaf's army then thundered forward. They painted their helmets with white crosses and marched to the war-cry *"Forward! Forward! Christmen! Crossmen! Kingsmen!"* The closer the armies got, the more they started cursing and shouting at each other, pumping themselves up for an unforgiving slaughter. Suddenly, the armies charged and combat began.

---

[21] Modern historians are sceptic to the *size* of the battle. Some claim it would not have been possible to assemble 14 000 men in that region, and that both armies were instead in their hundreds. Nevertheless, Olaf's was still outnumbered by a significant margin.

The men battled with grim determination, thrusting spears and hurling axes at each other. Olaf were in the thick of the battle, fighting on the front line with his axe *Hel*. The sight of the King fighting so valiantly motivated his men, who pushed hard into the enemy line. One of Olaf's skalds sung: *"Olaf was brave beyond all doubt, at Stiklestad was none so stout; Spattered with blood, the king, unsparing; cheered on his men with deed and daring."*

Meanwhile, Dag Ringsson applied much pressure on the enemy's left flank. Dag's men ripped into their ranks, inflicting much slaughter. Dismayed by such casualties, the farmers began to flee, surrendering the left flank to Dag. But in this crucial moment, when Dag should've swung at the centre to outflank the rest of the farmers army, he lost control of his men, who pursued the routing enemy instead of keeping formation. This was the fatal blunder of the battle.

At the centre, combat was full of havoc. Both sides suffered painful casualties, but neither would give ground to the other. Bent on slaying the King, Tore the Hound and some of his huscarls made their way towards Olaf's banner and, having found him, attacked. Olaf and Tore exchanged blows, until Tore managed to strike the King so he fell. One of Olaf's huscarls then hewed his axe at Tore's shoulders, wounding him momentarily, but Tore swiftly grabbed a spear and struck it right through the huscarl. As he got up, he saw Olaf fighting with Kalf, and immediately thrusted his lance at the King, piercing him between the legs and up his belly. Olaf fell on a large stone behind him and breathed heavily. Kalf then hurled a blade on Olaf's neck. He lay on the ground convulsing from the wounds, and mumbled a prayer, before passing.

With the sight of the King dead, Olaf's men lost all morale. Their line collapsed and the army broke apart, piece by piece. Dag Ringsson then finally came to attack the centre, but it was far too late. Olaf's army had already routed. Dag's men fought for a little while but were easily repulsed. The battle was lost.

Sainthood

It must have been quite a sight to see the cold fields of Stiklestad at the battle's end. The bloodstained ground was covered with dead men. Tore the Hound roamed the field in search of Olaf's body, hoping to have one last look at the old King he had helped in killing. But once he found him, he was struck by the untainted appearance of the corpse. As some of Olaf's blood touched his fingers, he became frightened and quickly left it. Tore claimed there was something holy about the body.

A few moments later, a nearby farmer named Torgil found the body, wrapped it and took it home. He and his son treated it, before laying it in a coffin and taking it to Nidaros. They presented it to bishop Sigurd of Nidaros, but after learning that Sigurd wished to dump the coffin in a fjord instead of giving it proper funeral services, they buried it themselves in all secrecy.

After the battle of Stiklestad, Cnut the Great installed his wife Alfiva, and son, Sweyn Cnutsson, as regents of Norway. But the time for Danish superiority was fading. Norwegians regarded the Danes as despots. Renewed resentment surged as many began to reminisce on their Kings. Maybe some even remembered the words of Olaf when he advocated Norway's right to independence. Those who had opposed Olaf began to regret their decision. Many searched for his grave to pay homage, but the grave was hidden.

What happened next has perplexed modern scholars: the people completely changed their opinion on Olaf Haraldsson. They began talking of Olaf as a holy man. Even his foremost enemies took a radical change in view: Tore the Hound affirmed that Olaf must have been a holy man and had already departed for a pilgrimage to Jerusalem. Einar Thambaskelfer also held this opinion. Even Kalf and other chiefs insisted so. Across Trondelag, common people *"began to say that King Olaf was truly a holy man."* Norwegian historian Ludvig Daae worded the puzzling phenomena accordingly: *"never has a people's shift in judgement been quicker or more profound than what now followed."*

Danish authorities soon demanded an explanation, so bishop Grimkell decided to exhume Olaf's grave, probably searching for any signs of holiness. Partaking in this was Einar Thambaskelfer and scores of other nobles. When they opened the coffin, they were left stunned. The body remained incorruptible. It had a beautiful, sweet scent and his cheeks were still red. His hair and fingernails had even grown. More people flocked around his grave to witness the miracle.

The Danish governess Alfiva was in denial: *"corpses decay slowly in sand. It would not have been like this if he lay in mould,"* she rebutted. She pledged to believe the incorruptibility if Olaf's hair did not catch fire. Grimkell cut a piece of the hair and put it in a sanctified lamp of burning incense – but the hair did not burn. Alfiva made further refutations, until Einar told her to shut up, giving her severe reproaches.[22] Sigvat the Skald witnessed this and composed a poem: *"I lie not, when I saw the king; seemed as*

---

[22] It should be noted that four *independent* sources from the 1500s AD state that Olaf's body lay incorrupt in its shrine, shedding an interesting light on the validity of Olaf's miracle.

alive in every thing; His nails, his yellow hair still growing; and round his ruddy cheek still flowing...the blind he cured he gave; a tress, their precious sight to save."

The body was buried again in St. Clement's church in Nidaros, next to the high altar. People flocked to the church from across the country, and word quickly spread of healings and miracles occurring by his grave. With the Danish governor's approval, Bishop Grimkell's acknowledgement and a popular vote at the Things, Olaf Haraldsson was officially declared a Holy Man. He became a sacred martyr for Norwegian independence and began to be called *Olav den heilage* – Olaf the Holy.

Yet, how can a brutal pirate warlord like Olaf become a Saint? A man who was fond of wealth, power and battle; who, in England, partook in gruelling acts against Christians and civilians; who suffered from an ill temper and punished enemies by cruel means? This question has led many Norwegian historians to discard Olaf's piety as some medieval fantasy or political trick.

However, we must recall a central element in Christian theology: a Saint was never supposed to be sinless. *No* human is sinless – that is a core principle in Christianity. The idea of the flawless Saint first began permeating European thought in the 18th century. Before that, it was common knowledge that all Saints had a tainted past, but whose genuine search for redemption ultimately gave them salvation through martyrdom. In other words, the *struggle for* and ultimate *reception of* grace was the focus, as it demonstrated the unlimited redemptive power of Christ. In the 13th century, the Franciscan theologian Bonaventure echoed this: *"Do you not know that many saints were sinful? When they committed great sins, they learned how to show mercy to us sinners."* Anyone, no matter how wretched, can find restoration in Christ if they are willing to face the humbling truth about themselves (self-denial), pick up their "crosses" and go through martyrdom.

What made Olaf a Saint, according to Catholic and Eastern Orthodox tradition, was his struggle to surrender fully to God, and his final success in doing so. All his life he had been among Vikings, so the Viking-mentality was all he knew. Yet, in his heart he strived to live as a Christian. A twilight. In the final year of his life, he was humbled and shattered as his people turned against him. Cnut robbed his Kingdom and he was reduced to nothing but a refugee. His encounter with the "glorious" man in Kiev marked the final surrender to Christ – the moment where Olaf submitted to his Creator's Will, which was to return to Norway and die fighting. A voluntary ride into death – the essence of martyrdom. His death at Stiklestad finally fulfilled his search for God: by submitting to God's Will, he could finally receive forgiveness and be cleansed of all his wickedness and sin. Throughout Olaf's Saga we hear explicitly of Olaf's sins, yet on the final pages we read profound words of contemporary poets, reflecting a different tone. As all men, he had sins – but Olaf found redemption in the end, and this was precisely the hope of every Viking convert who regretted their past crimes.

All this is reflected in the way St. Olaf is depicted in art. In statues, sculptures, religious and renaissance paintings, Olaf is found standing on a dragon or serpent. This is common in religious imagery, as a dragon usually symbolises evil or malign passions – but in St. Olaf's depiction, the dragon has the same head as Olaf's. This captures all that St. Olaf represents: being able to conquer one's own evil. His Sainthood reminded Christians of how Christ triumphs over anger, lust, greed, and so on; and therefore how a person who used to be a slave to all these passions – even to Olaf Haraldsson's extent – was absolved from them when he finally, and wholeheartedly, humbled himself before Christ through repentance. Through Christ, Olaf triumphed over his own dragon – his own evil.

He therefore became the conceptual *bridge* between the Norse world and Christendom. He showed that it was possible for even dreadful Vikings to find redemption. Christianity was an endlessly profound, zealous faith that had poured even into the heart of Olaf the Stout! And if such a terrible warlord as Olaf could receive it, why shouldn't other offenders? Furthermore, Olaf's post-mortem miracles raised the respect Norwegians had of Christianity. This helped the Church take roots, and, of course, helped the missionaries spread the Scriptures. Again, in Christian thinking, all this is not due to Olaf's works or deeds – but due to how God used Olaf's bones to work miracles.

The *Olaf-veneration* transcended Norwegian borders. So-called *Olaf-Churches* were erected across Northern Europe: on Ireland, in England, in Holland, throughout Scandinavia, in Germany and in Russia. His relics were placed in the Nidaros Cathedral, which soon became a popular pilgrim destination. People from all Northern Europe came to the field at Stiklestad to behold the battleground and the Nidaros Cathedral to pray by his relics. This very cathedral is today the northernmost cathedral in the world.

In Norway, he became the patron Saint of the country and the role model for all Kings and leaders. His Sainthood stimulated a literary and artistic flourishment during the High Middle Ages, producing some of the most beautiful songs, poems, statues, artwork and paintings in Norwegian history. Lars Roar Langslet, former Minister of Culture in Norway, wrote that he became *"a stimulating force of national integration. For many centuries, St. Olaf was a symbol for state and justice."*

The veneration of St. Olaf declined during the Protestant Reformation, which dismantled and looted many Catholic and Orthodox Olaf-churches. When the Danes turned Norway into a province of their own in the 16th century, they strived to eradicate his name, as he was a potent source of patriotism. He embodied the soul of the nation itself.

Legacy

"For fifteen winters o'er the land, King Olaf held the chief command; Before he fell up in the North, his fall made known to us his worth; No worthier prince before his day, in our Northern Land e'er held the sway; too short he held it for our good, all men wish now that he had stood."

– Sigvat the Skald, Olaf's loyal marshal, and court poet.

Beyond Sainthood, Olaf Haraldsson had an enormous impact on the Kingdom of Norway. He was the King that brought Norway into the feudal, European world. He restructured the country by creating a stronger, central government, helping to remove tribalism and promote national unity.

The strong autonomy and influence of Earls, chiefs and petty kings was arguably problematic because it maintained ancestral rivalries over the Kingdom's needs. Men of power acted not in the country's best interest, but in their own. As we have seen, this gave the Danes many opportunities to gain power in Norway. By deviously exploiting Norway's tribal politics, they expanded their holdings and occasionally made it their vassal state. They would take sides in the infighting, playing on the interest of various chiefs to tie Norway under their sway. As opposed to Norway, Denmark was a united, people under a single King.

Thus, Olaf broke down the old system, replacing it with a new model to unify the Norwegians. It would bind them to one King, one Church, under one God. In this sense, Olaf Haraldsson was a political revolutionary. But his revolution ultimately cost him the throne, and life. Uprooting of the old ways sparked fury among the nobles, who lost much of their wealth and status. The battle of Stiklestad was not a battle between "Christians and heathens", as often portrayed. It was a battle between the Old Norse and the European ideologies. Tore the Hound, for example, was a Christian who went on pilgrimage to Jerusalem after the battle. Kalf Arnesson, too, was a Christian. The feud they had with Olaf held political roots, not religious ones. His controversial ideas alienated him from his subordinates and had him killed. Only in retrospect did Norwegians realise the value of a more centralised, unified Kingdom.

A part of this revolution was the *formal* conversion of Norwegians to Christendom. I emphasise its formality. By setting up an institutional Church, Christendom was embedded into the core of Norway's state structure. Together with the King, the Church now lay at the heart of society.

The Christian Law *(Kristenretten)* had major influence too. Though Norse customs continued, they gradually lost relevance and ultimately, disappeared. St. Olaf is therefore a colossal factor in ending the notorious Viking Age.

Furthermore, there is reason to believe he contributed to the spread of Spiritual Christianity as well. The story of 1029-1030, Olaf's last living year, was indeed something special. It transformed the aggressive warrior into *a martyr*. His iconic death boosted the spread of Christianity, as it unleashed an unprecedented curiosity and admiration for the Christian way of life. This flaming interest created a Christian Spiritual revival that solidified Christianity as Norway's official religion. Interestingly, it also helped realise and bolster his vision - a unified Kingdom built on Christian values. *His very death revolutionised Norway.*

We may therefore recall the anecdote of how Olaf's first step on Norwegian soil made him trip and fall, and how his huscarl quipped *"Now you gained your foothold in the land!"* We realise that this remark would come to be both prophetic and symbolic. It sums up what made Olaf special: his fall won him a lasting foothold in Norway.

St. Olaf's display of resolution, courage and devotion to his ideals inspired many contemporary and later Norwegians alike. His sacrifice for his country served as a source of motivation during Norway's defining moments, like the declaration of independence in 1814 and the establishment of modern Norway in 1905.

Even in the dark years of the Second World War, St. Olaf's everlasting name dwelled amongst the people. In 1940, King Haakon 7th of Norway refused to relinquish his throne to the invading Nazis and was instead forced into exile. This was a disheartening moment in Norwegian history, yet the memory of King Olaf emboldened the people. Carl J. Hambro, Norway's Parliamentary President during the war, ended his heartfelt memoir *I Saw It Happen In Norway* (1940) with these words:

"There was another King of Norway who had to go into exile with his son, the Crown-Prince. That was 911 years ago. The King's name was Olav Haraldsson... King Olav went to Russia with a small body of followers, and he had no allies. He was offered the kingdom of Great Bulgaria, which he refused, and it is stated in the Saga that he had it in mind to abdicate formally, especially after his enemies tried to raise the people against him. It is told in the Saga that he was most sorrowful and he asked God to be his judge; he turned things

over in his mind and did not know what to choose; it came to him that any decision might be unfortunate.

It is related: 'One night King Olav lay in his bed thinking over his plans, and his mind was filled with great sorrow. But when his spirit was very tired, sleep came to him, but a sleep so light that he seemed to be awake and see everything that happened in the house. He saw a man standing in front of his bed, tall and glorious and beautifully dressed... The man addressed him: '... It seems strange that you are wavering to and fro and still more that you can imagine relinquishing your Kingship, given to you by God... Don't let your inferiors scare you. It is the honour of a King to defeat his enemies, but an honourable death to fall fighting with your men. Are you in doubt whether the right is yours in this struggle? Never deny your true right. You will brave the dangers and find Norway, and God will bear you witness that it belongs to you.'

When King Olav woke up he had made his decision. He came back to Norway. He became Saint Olav, eternal King of Norway. And his descendant, King Haakon [7th], is occupying his throne to-day."

Magnus the Good

1024 – 1047

"Trust not him whose father, brother or other kin you have slain, no matter how young he be. For often grows the wolf in the child."

– Volsunga Saga c.21.

Truly, this young man was a remarkable character. Under his reign, the tables turned completely. Norway became a Northern giant. It was a time when Norwegians, inspired by St. Olaf, desired their own powerful realm – to stand independently and mightily. Magnus and his companions catapulted the country into a new age of military aggressiveness and national rise.

The Country that Needed a King

"I was no friend of Olaf. But the Tronds were bad tradesfolk when they sold their King in exchange for this bit** and her companions. This king [Sweyn] cannot talk, and his mother [Alfiva] wishes us only evil." Such were the alleged words of Earl Einar Thambaskelfer, spoken at a large thing in Trondelag in front of vice-king Sweyn and his mother Alfiva. Einar represented the common sentiment among the people. They felt oppressed. The Danes had imposed high taxes and implemented a set of tyrannical laws, perhaps with the hopes of crushing the Norwegian growing spirit of independence. Even the Norwegian earls resented their Danish overlords. Their promised privileges, wealth and honours never appeared. Their wishes and demands, neglected. Einar Thambaskelfer and Kalf Arnesson were particularly disappointed. They began advocating for Norway's independence from Denmark and entitlement to its own King.

However, the Norwegians had just killed their King. Therefore, there was only one way for a King to rise again: through St. Olaf's only son, Magnus, who had been left behind in Novgorod. Though Magnus was born out of wedlock, he was their only hope.

In 1034, Einar and Kalf left Norway, sailed across the Baltic, galloped through the Estonian woods and to Novgorod at all haste. Once there, they met St. Olaf's exiled family and, most importantly, the little boy Magnus. Kalf and Einar pleaded the family for permission to take the lad and make him King, emphasising the unbearable tyranny of the Danes and the sanctity of the Norwegian royal throne. After a sacred oath never to betray the boy, their request was granted.

Kalf and Einar recruited Rus' soldiers along the way. When they arrived in Sweden, King Anund Jacob also sponsored them with hundreds of able men. To add to their luck, Cnut the Great died unexpectedly in 1035, leaving the Danish grip on Norway in jeopardy. Moreover, Cnut's North Sea Empire was divided among his two sons: Harthacnut ruled Denmark and Harald Harefoot ruled England. The two successors lacked the verve and talent of their father, and it was possible that the North Sea Empire could crumble. Encouraged by such news, Kalf, Einar and Magnus crossed into Trondelag during Autumn of 1035. A major thing was held there, where Magnus, thanks to his father's outstanding legacy and the rotten infamy of the Danes, was hailed unanimously as King of Norway. The boy was eleven years old.

In Trondelag, Magnus made personal encounter with a woman in her late 30ies. The life of this woman had been tough so far. Cynical soldiers had mocked and harassed her for years, especially after the death of her lover and master. Magnus did not recognise her at first, but then she revealed her identity. She was Alfhild, Magnus mother and St. Olaf's former concubine. Emotions must have run deep. Magnus embraced her. He would protect her dearly for the rest of his life.

The young prince also visited his father. He ensured to have the body of St. Olaf placed into a golden sarcophagus, rich in ornaments and jewels. Magnus had few memories of Olaf, but he heard plenty of stories from other people, who now venerated him as a holy man and a glorious King. Magnus must have sensed the pressurising expectations people had of him as Olaf's son. Could he become as successful as his legendary father?

Sweyn and Alfiva heard of what was unfolding, but they could do nothing. No free man was willing to fight for them any longer. Sweyn even tried to receive troops from Denmark, but Harthacnut denied him. Harthacnut was more concerned of his Anglo-Saxon inheritance than the rebellion in Norway. Therefore, by Christmas 1035, Norway officially had its undisputed, new ruler: the young Magnus Olafsson.

Treaty at Göta river

As long as Magnus remained a child, Einar and Kalf ruled unconditionally through him. They certainly took advantage of this opportunity. It was time for their vision to be realised. They abolished many of the laws from Sweyn and Alfiva and added new laws to decentralise the monarchical power and restore the pre-Olaf, Norse system. They would slowly dismantle Olaf's power structure as a return to the old ways. Times were good for the earls – but it lasted only while Magnus stayed childish.

In 1037, the uneasy relationship with Denmark escalated. The Danes invaded Viken and Ostfold, but were met by Magnus' army, led by Einar and the earls, and subsequently beaten. The Danes withdrew to the Göta river to offer battle, but instead, nobles from both sides began peace talks. Though Magnus and Harthacnut were both present, they were too childish to participate (Magnus was 13 years old and Harthacnut was 17), so the negotiations were instead conducted by their caretakers and court advisers.

In the end, they had devised a rather peculiar arrangement: to settle the disputes and guarantee peace, Magnus and Harthacnut would become each other's heirs. The one who died first was bound by oath to cede all his lands to the other. They swore to never wage war on each other, and twelve principal nobles on each side swore that they would honour this treaty as long as they lived. It was now a matter of staying alive – to outlast the other.

Vengeance

Magnus Olafsson matured fast. Witnessing the horrors of Harthacnut's raiding in Viken and Ostfold had made a serious impression on him. He became more conscious about the life and death of his father, especially when former huscarls of St. Olaf openly criticised him for dining with his fathers' murderers.

His cup ran over at the age of 16, when he asked Einar to ride with him to Stiklestad and give a detailed account of what had happened there. Einar shrewdly answered that he knew nothing of the battle, for he was not present. Kalf, however, was there and should ride with the King, Einar deviously suggested. A startled Kalf immediately excused himself: *"That is really not my duty."* But Magnus snapped: *"Go thou shalt, Kalf!"* They rode together to the battleground, and when they arrived, Magnus asked: *"Where is the spot at which the King fell?"* Kalf showed him. Magnus then asked the dreaded question: *"Where were you then, Kalf?" "Where I now stand,"* a wary Kalf replied. The young King stepped forth and stared his caretaker in the face: *"Then thy axe could well have reached him."* Knowing he was exposed, Kalf rebutted *"my axe did not near him!",* mounted his horse and rode away at all haste. It was clear that little Magnus was not so little anymore. Kalf fled the country.

Magnus became determined to avenge his father. He commenced an uncompromising hunt to wipe out all who had conspired against St. Olaf. Royal huscarls stormed from household to household, punishing Olaf's former enemies. Horic of Tjøtta saw his end here, as he was axed to death. Many were expelled and had their estates confiscated. Others were directly put to death. Some were mutilated. Magnus showed no mercy.

Witnessing all this was Sigvat the Skald, the former court poet and marshal of St. Olaf. Sigvat was Magnus' godfather. He was also the one who had given him the foreign name *Magnus,* after Charlemagne (*Carlo Magnus* in Latin)[23]. As Magnus grew up, Sigvat maintained a close relationship with him. However, he was now deeply concerned by what he saw in the young King. Though Sigvat advocated justice, he criticised Magnus' excessive brutality. He warned that it would revive divisions and old feuds – threatening to tear the realm apart.

During a feast, Sigvat suddenly stood up and performed his poem *A Free Speaking Song.* It opened with a reminder of Norway's inheritance from Haakon the Good and the two Olafs, and how these Kings never broke their promises. *"Thy honour lies in thy good sword, but still more in thy royal word,"* Sigvat advised. *"Dreaded King! Who urges thee to break thy pledged word?...the battle-storm raiser he, must by his own men trusted be. Who urges thee, who seeks't renown, the farmer's cattle to cut down? No king before e'er took in hand, such Viking work in his own land. Such rapine men will not long bear... when once inflamed, the king himself for all is blamed."* After further exhorting the King to obey the laws of his fathers and end this murderous campaign, he finished the song with these words: *"a holy bond between us still, makes me wish speedy end to ill... with thee we'll fight by sea or land, with Olaf's sword take Olaf's mind, and to thy farmers be more kind."*

Upon finishing, royal huscarls apprehended him, but King Magnus immediately ordered them to stand down. Magnus loved Sigvat and had listened very carefully. He decided to take his counsel and end the campaign of vengeance. Poetry had tamed the vindictive King.

---

[23] According to the sagas, Olaf Haraldsson was sleeping when his concubine, Alfhild, gave birth. Sigvat was present and helped her. Olaf had strictly told his subordinates not to disturb his sleep, so Sigvat resolved to quickly baptise the infant in case the infant would die unexpectedly (before baptism) and named him Magnus. On the following day, Olaf asked Sigvat why he had given the boy such a strange, foreign name – as probably no one in Norway were called *Magnus* at that time – to which Sigvat answered: *"I called him after King Carlo Magnus, who, I knew, had been the best King in the world."*

Sigvat's poem couldn't have been more timely. Norway was at the brink of renewed civil war: Norwegians from Sogn had already mustered for war against the King. But as the situation settled, they put down their arms.

Magnus became very interested in the law and is credited for putting the great Frostathing legal code into writing to preserve it. He would go on to rule Norway with strict adherence to laws and customs and earned a reputation for being noble and chivalrous. His godfather and mentor, Sigvat the Skald, died peacefully in 1045.

The Rise of Norway

The young King became increasingly impatient over Harthacnut, who, following the death of Harald Harefoot, now ruled Denmark *and* England. If Harthacnut died, Magnus could now lay claim on both countries. Guiding him in all this was the clever Einar Thambaskelfer. Einar had survived Magnus' campaign of vengeance, having convinced Magnus that he had never been a serious enemy of his father (although Einar had sided with Olaf's enemies, he never personally feuded with him). Einar was 60 years old and highly competent as a military leader. His career alone was impressive: remember, Einar fought alongside Olaf Tryggvason at Svolder in year 1000.

Magnus and Einar consulted Sweyn Ulfsson Estridsson, the son of the mighty Earl Ulf. The Danes had expelled his family and killed his father due internal complications. This had embittered Sweyn and made him covet for revenge. But Sweyn's treacherous intensions were unknown to Harthacnut, who instead trusted him with governorship of Denmark while he was in England. In June 1042, Harthacnut drank poison and died, giving Magnus a legal claim to Denmark and England via their treaty. Magnus and Einar immediately mobilised an army and, with Sweyn's help, invaded Denmark.

Surprisingly, the Danish people offered little resistance and instead rallied behind Magnus, whom they perceived as generous and noble, as opposed to Harthacnut, whom they despised for his harshness. Besides, Magnus had a perfectly lawful claim to the throne. All treaties ought to be honoured, so the Danes saw no reason to resist him.

Magnus assumed the throne of Denmark unhindered. He then had the Danish nobles swear an oath of allegiance, rewarding them with fiefs in return. He also organised the land into baronies and districts to make it easier to administer. Sweyn Estridsson remained the appointed Earl of Denmark – a decision Einar deeply disagreed with. Einar felt that Sweyn could not be trusted with such an important position. If Sweyn could betray Harthacnut, he could betray Magnus. *"Too great an Earl, too great an Earl, foster-*

son," he warned Magnus. However, Magnus sought to wrench himself from Einar's influence and ignored his warning.

By the end of the year 1042, just twelve years after the death of St. Olaf, Magnus Olafsson was the King of Norway *and* Denmark. At the age of 18, he was already more powerful than his father had ever been. The tables had turned completely.

The Wendish Campaign and the Battle of Lyrskov Heath

Stepping in as Denmark's ruler was not an easy task. By doing so, he was a successor of the great Cnut himself. Many issues were unresolved and needed attention. Yet, Magnus aimed higher. He had ambitions of reclaiming England, his second half of Harthacnut's inheritance. If he succeeded, he would restore Cnut's North Sea Empire.

For the moment however, England was lost. Harthacnut failed to consolidate Nordic authority there, and his premature death allowed Anglo-Saxon nobles to reassume control. Most active in these affairs were Earl Godwin, who helped Edward the Confessor, Ethelred's son, to become England's King.

Magnus sent a dire letter to them, emphasising his legitimacy to the throne and demanding Edward's abdication, threating a colossal Nordic invasion if otherwise. Edward replied that his dignity and honour as King would not permit him to abdicate the throne voluntarily. *"He shall only get the opportunity of taking England when he has taken my life,"* Edward reportedly said. Anticipating a Norwegian invasion, Edward moved his troops to England's eastern coast.

Yet before Magnus could plan this grand invasion, he had to secure Denmark. The land was under threat by the Polish Kingdom of Wendland. The Wends had, as historian Kim Hjardar holds, seized the citadel Jomsborg – wiping out the renowned Joms Vikings, a people that previously served as Denmark's close allies and main military muscle. They now ravaged Danish settlements on the border and threatened invasion. It was time for Magnus to step up and protect his people.

In 1043, Magnus assembled his army and assaulted Jomsborg – a fortress with 10 000 inhabitants. After penetrating the walls, he let loose his dogs of war: the men stormed the citadel and butchered everyone in their path, even civilians. They pressed on to the inner part of the fortress, where the Wends had rallied and entrenched themselves. There was high ground here, making it very hard for Magnus' men to breach it. Finding no way into Jomborg's core, they retreated. However, this does not mean they lost. By devastating the fortress of Jomsborg they had successfully retaliated and weakened the

Wends. What was once a source of military might, was now a gloomy ruin. Magnus engaged with the enemy another two times later that year, with indecisive results.

In September, the King received reconnaissance of a huge Wendish army approaching. If Magnus withdrew, he would lose his momentum, demoralise his army and lose honour among his subjects. He had no other choice than to face the superior enemy. Luckily, in this hour of need, the Holy Roman Empire came to the rescue. They were eager to crush the pagan Wends, who had conducted heavy persecutions of Christians and torched churches in the region. A battalion led by the German Count Ordulf joined Magnus' men, and with a combined Norwegian-German army, they marched forward to meet the Wends head-on. They met near Lyrskov Heath, about 6km outside Sleswig in Germany.

Magnus' army soon lost heart when they laid their eyes on the staggering size of the Wendish army. Sources claim they were outnumbered by at least 15 000 men. Seeing the wariness of his men, Magnus allegedly threw away his chainmail and raised his father's legendary axe, *Hel,* and stormed alone towards the Wends. Seeing this, the entire army found courage and charged with him. They smashed into the Wendish lines and stormed through them with stunning ferocity. The Wends were completely unprepared for this shocking attack. They probably expected Magnus to take a defensive position, as he had inferior numbers. The Wends probably didn't even have time to organise into a defensive position, and their entire army was in complete disarray. Chaos spread, and with it, fear. Their lines broke completely, and, after just a short period, they began to rout. Magnus' army followed in a hasty pursuit and cut down as many as possible. They took no prisoners. The Wendish army had been completely overrun.

The Wends sustained enormous casualties, as most of their men fell during the rout. Those who survived deserted. The army was virtually annihilated. Magnus' casualties consisted of a few deaths, but many wounded. Magnus personally visited his wounded and helped treat them. In the aftermath, the Danes flocked around Magnus and raised him up, praising his name: *"Magnus the Good!" "Magnus the Good!" "Magnus the Good!"*

Power struggles

King Magnus returned triumphantly to Denmark and was applauded by Danish nobles. He had won the deep respect of the Danish people. Watching this envelop was Sweyn Estridsson, who enviously plotted to overthrow the young Norwegian and seize Denmark for himself. He secretly contacted Swedish King Anund Jacob, who was concerned over the sudden presence of a strong Norwegian realm. With Swedish backing, Sweyn began his revolt. Denmark was thrown into a civil war.

For Magnus, such news couldn't have been more alarming. It not only postponed his plans for England, but it caused deep unrest among the Danish people, jeopardising domestic stability. Hoping to crush Sweyn quickly, he hurriedly raised an army and attacked Sweyn at once. They would go on to fight three pitched battles, all of which Magnus won. He then ravaged and ransacked all provinces that offered men and resources to Sweyn. Some provinces were severely terrorised.

With Sweyn expelled, it was time to resume the mobilisation for England – but again, such plans were postponed. Harald Sigurdsson, Magnus' uncle, returned from his military service in Constantinople and now demanded his share of the royal inheritance. He brought with him tons of gold and hundreds of men from the Far East. To lay pressure on Magnus, Harald allied with Sweyn Estridsson and raided in Denmark – putting serious strains on Magnus' treasury.

Magnus called for aid from his mentor Earl Einar, Norway's governor. Einar and Magnus campaigned in Norway, cautiously pursuing Harald but avoiding direct confrontation. They feared Harald Sigurdsson, for he was a renowned tactician and warrior. Not wanting to gamble by war, they decided to resolve the situation diplomatically.

The meeting between them was tense at first. Harald was particularly suspicious of Earl Einar, who kept whispering counsel in Magnus' ear. *"You are high and mighty now, Einar, and now as always you are making your opposition to me clear. It will be a fine day when pride takes a fall. By as much as you are now a head taller than everyone else, you will soon be head shorter,"* he threatened.

However, after a few exchanges, the discourse gained a friendlier tone, especially between Magnus and Harald (who were, after all, nephew and uncle). In a final offer, Magnus offered to share his throne in exchange for half of Harald's treasures. Harald would be king of Norway, but remain *second* to Magnus. Harald agreed, and the two formed a new, amiable relationship (which was probably concerning for Einar). With this new treaty, Norway-Denmark was split, though ruled by the same dynasty.

Having dealt with all troubles in Scandinavia, Magnus' hands were free to tackle England. Having obtained half of Harald's enormous wealth, he had plenty of funds to finance an army and fleet.

However, it would never happen. Magnus the Good died in an accident in 1047. Sources disagree over what kind of accident it was. Some claim he fell off his horse, landing on a rock that incurred a fatal injury. Others that he slipped while embarking a boat, falling into the sea and drowning. Was it an assassination? Regardless, Magnus "the Good" Olafsson died. He was just 25 years old.

This was a tragedy. The entire people mourned his passing. During his grand funeral in Nidaros, all the bells rang and thousands of people flocked around his casket to pay their respect. Even the toughest of men wept. The sagas of *Morkinskinna* tell us: *"Everyone said that no man had been so dearly loved by all the people in Norway as King Magnus."* He was buried next to his father.

Legacy

> *"The tears o'er good King Magnus' bier, The people's tears, were all sincere: Even they to whom he riches gave, Carried him heavily to the grave. All hearts were struck at the king's end, His house-slaves wept as for a friend; His court-men oft alone would muse, as pondering unthought of news,"*
>
> – Od Kikinskald, a court poet.

Magnus the Good was very conscious over his father's career and legacy. Ever since he visited his fathers' grave in 1035, he obliged himself to honour his father in the best way he could. At first, he naively shared power with the Earls, but later, he turned into a serious avenger.

Driven by the expectations people had of him as St. Olaf's son, Magnus' ambitions knew no bounds. He became a relentless martial King who loved his huscarls, and they him. Magnus rewarded his soldiers so richly that he almost went bankrupt at the end of his reign.

Magnus the Good displayed a remarkable leadership talent, with a quick wit and strategic savvy. But it is important to emphasise that his leadership success owes credit to the men who tutored and advised him: Sigvat the Skald, the godfather who kept him from ruining both himself and Norway with his obsession for vengeance; Einar Thambaskelfer, who had placed him on the throne, co-devised the important Treaty of Göta river and advised him on wars, strategy and geopolitics.

Yet, once on the throne, Magnus carved his own, extraordinary path. His triumph at Lyrskov Heath echoed across Europe, gaining him respect among all Nordic peoples.

Many focus on the bitterness of his unfortunate death – but let us rather focus on his achievements while he still lived.

Magnus reign represented a new era in Norwegian history: an era when Norway finally stood united. Through St. Olaf's revolution, the Treaty of Göta river and Magnus' successes, the Fairhair dynasty had expanded its claims beyond Norway's borders. The dynasty was now a rightful inheritor of Cnut the Great. This meant they were destined to restore the North Sea Empire. It meant they were destined to take back England.

Harald Hardrada

1015 – 1066

"Our army's wing, where I shall stand. I will hold good with heart and hand. My mother's eye shall joy to see, a battered, blood-stained shield from me. The brisk young skald should surely go, into the fray, give blow for blow. Cheers on his men, gain inch by inch, and from the spear-point never flinch."

These were the words composed by a young Harald Sigurdsson, shortly after the Battle of Stiklestad. The words echo the pride, boldness and adventurism Harald embodied. His wanderlust would take him to the far places of the world, and his thirst for power would change the course of history. He was a true international – from serving the Greek Emperor in the Middle East to fighting valiantly at Stamford Bridge in a year that forever changed British, and indeed world, history. His life story is as extraordinary as it is important.

First blood at Stiklestad

Harald Sigurdsson was the son of Aasta (Åsta), St. Olaf's mother, and Sigurd the Sower (Sigurd Syr). He was therefore King Olaf's half-brother. One day, King Olaf came to visit his parents and met little Harald, who was just three years old at the time. He saw the boy play with small wooden warships and began talking with him. Charmed by the unusual wit of the child, Olaf turned to his mother and declared: *"Here, mother, you are bringing up a King."*

Harald looked up to Olaf. He showed interest in war already as a young boy. An opportunity to experience the rush of war presented itself in 1030 AD, when his brother Olaf returned to Norway to reclaim the throne. Harald was only 15 years old, but he begged Olaf for permission to join him in battle. Olaf initially denied him, to which Harald reportedly insisted: *"Certainly I shall be in the battle! For I am not so weak that I cannot handle the sword; and if I am, then my hand should be tied to the sword-handle. No one has a greater will to do the farmers more damage than I."* He got his will.

But the battle of Stiklestad was a carnage that Harald would never forget. The combat was terrible and traumatic. Suddenly, his beloved brother Olaf lay slain on the field and panic spread. Harald was lost in the chaos and was suddenly struck down and severely wounded. He was dragged away the battlefield by a huscarl, who took him to a nearby farm where he was treated for his wounds. Having regained just enough strength to stand up straight, Harald took a horse and rode away, fearful of being caught by the enemy. He perished into the dark woods, exhausted, alone and wounded. He composed these lines: *"my wounds were bleeding as I rode; and down below the enemies strode; killing the wounded with the sword, the followers of their rightful lord; from wood to wood I crept along, unnoticed by the enemy-throng; 'who knows' I thought, 'a day may come; my name will yet be great at home?'"*

Harald crossed into Sweden, where he regrouped with other survivors from the battle. Together, they rode through Sweden and into Russia, riding all the way to Kiev, where Grand-Prince Yaroslav the Wise stood ready to greet them. Harald burst into the courtroom, fatigued from the long journey. He met a six-year-old Magnus Olafsson there, who soon learned the fate of his father.

Harald agreed to enter military service for the Kievan Rus. The year was 1032 and Harald probably participated in the Kievan expeditions against the Pechenegs around the Black Sea. Due to his royal blood, he quickly became a leading commander in the army.

He also befriended and fell in love with Yaroslav's daughter, Elizaveta. He asked Yaroslav to grant him her hand in marriage, but to Harald's surprise, Yaroslav rejected him. he had yet to prove himself, Yaroslav affirmed.

This made Harald consider more ambitious ideas – he decided to voyage to *Miklagard*, the Norse name for Constantinople, meaning *the Great City*. Here he could fight for the Eastern Roman Empire[24] and make a fortune. Ahead of 500 warriors loyal to him, he bid farewell to Yaroslav and sailed South.

A Wolf in the Orient

Scandinavians had often travelled to Constantinople to serve as members of the Emperor's Imperial guard, the so-called *Varangian Guard*. The Varangian Guard was

---

[24] In Harald's time, folk referred to the Byzantine Empire as the Eastern Roman Empire. The term "Byzantine" is a rather modern creation. The Norsemen also called it *Grekerriket*, meaning *the Greek Empire*.

the imperial legion of elite warriors in the Byzantine Empire, composed mainly of Scandinavian and Russian Vikings. Its origins can be traced back to a contract between the Rus and the Greek Emperors, where the former promised to supply Norse warriors to the latter. But the formal establishment of the Guard did not happen until the end of the 900s, when Emperor Basileiros deployed Norse merchants in a battle versus Bulgarians. Overwhelmed by the Bulgarians, all the emperor's infantry fled – except for the Norsemen.

In the end, the Norsemen managed to exhaust the Bulgarians and force them to flee, ultimately winning the battle. So impressed was Basileiros that he decided to recruit more of these Scandinavians and use them as his personal guard. Ever since, thousands of Scandinavians had voyaged to *Grekerriket* ("The Greek Empire") and been recruited as elite soldiers for the Emperor. The term "Varangian" derived from the Old Norse word *varya,* meaning oath or pledge. *Variangians* therefore means *"men of the pledge"* or *"men of oath."*

In 1034, Harald arrived and applied to join the Guard, but the Byzantines first deployed him in the Navy. His first mission took him to the Levant, hunting and fighting Arabic pirates. His unit first engaged with them at sea, but later, in Asia Minor. He joined the campaigns that pushed the Arabs all the way to the river Euphrates.

According to medieval writings, a staggering eighty Arab fortresses succumbed to Harald and his men. He also fought in the Holy Land, allegedly cleansing the area from brigands and thieves. It is believable that he visited the Church of the Holy Sepulchre in Jerusalem and bathed in the river Jordan. Having served with distinction on numerous occasions, Harald was most likely admitted to the Varangian Guard by this point. He was already building himself quite a reputation and assumed a high rank.

In 1038, Harald was deployed to Sicily as one of the marshals of the famous Byzantine General George Maniakes to fight the Saracens. Harald appear frequently in the Byzantine source *Advice to the Emperor* under the name of *Araltes* – a ruthless and unstoppable "berserker".

The wealthy towns of Sicily underwent severe pillaging from Harald and his troops. Some Greek sources report of several atrocities as Harald sought to spread fear and terror in the hearts of the enemy. He linked up with other Normans in the area, like William Iron-Arm, Drogo, and Humphrey, and fought together with them. The Normans, of course, originally stemmed from Norway and Denmark, and could probably still speak the Norse language. Harald probably regarded them as his kinsfolk.

Sagas claim that Harald and his company conquered four towns, numerous fortresses – including one that had never been breached before - and won eighteen battles. They overcame robust defences often by ingenuity. One time, Harald had his men capture dozens of pigeons, soaked them in candle wax, lit them on fire and send them back behind the walls where they had their lairs. With the fortress now an inferno, Harald and his men easily scaled the walls and conquered. Another story claim Harald once pretended to be ill and dead. His men politely asked the townsmen if they could bury him properly in the town. Once inside, a 'resurrected' Harald sprung up from his coffin and, helped by two Varangians, kept the gates open for the rest of the army to enter and storm the town.[25] "[Harald] *dyed the eagle with bloody specks. And where you raided, the wolves were fed,*" the skalds sung about him. Observing how wolves and vultures fed on the corpses of his enemies, Harald soon received the horrifying tags *"the wolf"* and *"the wolf-feeder."*

All the while, Harald amassed great riches. The origins of his wealth were many: riches seized from pirates, gifts from friends and subjects or plundered loot from Sicily and Africa. He deliberately underreported the size of his loot and smuggled these riches to Kiev for safekeeping. However, General Maniakes already suspected Harald for embezzlement and was determined to arrest him.

Emperor Michael IV favoured Harald and frequently promoted him. The two spent good time together, giving Harald access to the Imperial court. In fact, Emperor Michael IV made him tax collector *and* a leading commander during the war against Bulgarian rebels. Harald led the troops in his usual style, earning him the name *Harald Bolgara Brennir,* meaning Harald the Bulgar-burner. The rebellion was crushed, and Harald returned to Constantinople alongside the Emperor in triumph. He was at the pinnacle of his Varangian career.

But when Emperor Michael IV died in 1041, the tide swung against Harald's favour. He was despised by the new regent, Empress Zoe. She was probably suspicious of his increasing involvement in Imperial affairs. His close relationship with the former Emperor had already raised concerns. Was this brute from Norway trying to gain Imperial power? To make matters worse, Harald allegedly had a romantic affair with Zoe's granddaughter, Maria.

Zoe could not just rid herself of Harald outright, for the Varangian Guard loved him. They needed him to be guilty of a crime – and one day, they got it. General

---

[25] More stories like this can be found in the sagas *Heimskringla, Fagerskinna* and *Morkinskinna.*

Maniakes publicly accused Harald of embezzlement and Harald was immediately arrested and imprisoned. In such a short time, he had gone from the grandiose days of Imperial service, to the dirty pits of the dungeon.

Luckily for Harald, the entire city was soon thrown into anarchy. After an attempt by Michael V to oust Zoe, a rebellion struck the capital. Soldiers stormed the palace and freed numerous political prisoners – one of them Harald. He joined the Varangians as they ravaged the Emperor's palace.

Harald probably took advantage of the situation, looting as much treasury as he could. Covered by the disarray, he slipped to the docks where he prepared two galleys and filled them with treasures and his remaining friends. They set sail but found themselves trapped as the Greeks had bound a chain between the gates of the port. Harald ordered his men to gather at the back of the ship, making it weigh lighter at the front so that the galley could slip over the chain. Once halfway, Harald and his men rushed to the front of the galley, making it tip over the chain. It worked for Harald's ship, but his other galley broke in half, sending the men and the treasures to the bottom of the sea. Nonetheless, Harald Sigurdsson successfully escaped Constantinople. His Oriental adventures had come to a heart-pounding end.

The Thunderbolt Returns

Harald returned to Kiev, now a living legend. When aggregating the riches he had deposited in Kiev over the years, it was clear that he was rich beyond comprehension. Grand-Prince Yaroslav was certainly satisfied to say the least. He finally gave his daughter, Elizaveta, to the Varangian prince, and they married. Harald loved Elizaveta dearly, and wished to make her Queen of all his possessions.

However, the newly-wed Harald would soon embark on another adventure. Norwegians in Kiev told him about the plight in Norway, about Magnus' rise to power and stunning victories in Denmark. Harald held that Magnus was too young and inexperienced to rule and viewed himself as a more promising ruler. Knowing he had a claim to the throne, he headed North. In 1045, Harald arrived back in Scandinavia. He had left it as a fugitive but returned a rich and mighty warlord.

War drains resources and leads only to destruction. An intelligent general knows this and achieves his goals without bloodshed. Harald therefore negotiated with Magnus instead of fighting him, and the two reached an agreement. Harald would cede half his Byzantine gold to Magnus in exchange for the throne of Norway. He essentially bought the throne in gold rather than manpower. With the treaty enacted, he returned to his homeland and became its King. His wife, Elizaveta, became Queen of Norway.

It was at this time that Harald Sigurdsson earned the name *Hardradi,* meaning Harsh-ruler. He perfectly remembered the fate of his half-brother Olaf and was determined to not suffer the same betrayal. He would strike hard at his conspirators before they could strike him. Doing so involved not only arrests, assassinations, or executions, but also a final consolidation of the feudal monarchy. As we know, under King Magnus, Einar Thambaskelfer had worked to resurrect the pre-Olaf, old Norse system that favoured the earls. With Harald's arrival, this was stopped as Einar's reforms were nullified. Harald disempowered the earls again. Einar realised he was in grave danger.

In the wake of Magnus the Good's death in 1047, public support of Harald took a sharp dive. Compared to their beloved Magnus, Harald appeared as an oppressive tyrant. Einar Thambaskelfer saw an opportunity to make his move. He fuelled public resentment by speaking on things against the King. *"To follow a dead Magnus is better than to follow any other living King!"* he shouted. To calm down the domestic unrest, Harald took a second wife in Tora Torbeigsdottir, a woman from a prestigious family in the Westlands and Lade[26] – but it had little effect. Einar continued to defy Harald's orders, and began gathering a rebel army. The King decided to put an end to this rebellious Earl once and for all. While pretending to seek reconciliation, he invited Einar in for a meeting. As soon as Einar entered, he was struck down and murdered by one of Harald's guards.

Einar's death infuriated the people and the earls even more, but Harald's iron fist struck hard and silenced all criticism. He was always one step ahead and trumped every revolt or opposition. Another revolt led by Earl Haakon Ivarsson, Einar's counterpart, was also quickly put down. Earl Haakon fled to Sweyn Ulfsson Estridsson, who at that point had become Denmark's new King.

Harald *the Landwaster*

Since Magnus' death, the Danish crown found its way gently on top of Sweyn's head, which Harald immediately disputed. Harald's and Sweyn's war opened with a series

---

[26] Tora became Harald's second wife and never became Queen of Norway. Harald's first wife however, Elizaveta, became the official Queen of Norway. Harald also wanted to make Elizaveta Queen of England.

of sporadic attacks on Denmark. Harald sought to ruin Sweyn's finances. His targets were mostly farms, villages, and towns, but he sometimes plundered very important cities. For instance, Hedeby –the trading jewel of Denmark – saw its bitter end at the hands of Harald's Norwegians. They not only sacked it, but also burnt it to the ground. Hedeby never recovered. We can understand why Harald's contemporaries called him *the Landwaster.*

All while this happened, Sweyn restlessly tried to stop his foe, but with no luck. His closest attempt was near Fyn in 1051, when Sweyn's armada stumbled upon Harald and a few of his ships. It didn't take a tactical mastermind to calculate that defeat was imminent, so Harald sounded retreat, but due to the heavy treasures he had stolen, his ships were too slow. Sensing incoming disaster, Harald ordered his men to toss out all the treasures to make the ships lighter. It accelerated them, but Sweyn's fleets were still faster. Now desperate, the Norwegians threw out all Danish slaves and prisoners. Finally, they were able to break loose and head home to safety.

The following years were characterised by smaller engagements, skirmishes, and raids in Denmark, but in 1062, Harald decided to seek a decisive battle. It took place at the Nis river (*Nisså*). Harald allegedly had 180 ships against Sweyn's 300. He tactfully opened the battle with archers, spraying the Danes with arrows for as long as possible. Then commenced the main, ghastly phase, as the ships rammed into each other and boarded for clearing. The battle seemed never-ending as it continued over midnight. Once deep into the night, Sweyn's ship was cleared and the Danish King fled. His leaderless fleet was destroyed.

Harald rode at all haste inland to hunt for Sweyn, but the Dane had disappeared. With the Sweyn alive, Harald knew the war would live on – and so it did. Further clashes continued the next two years. Although the Norwegians won every battle, they failed to secure a decisive result.

We must remember that, contrary to Magnus, Harald was infamous for his harshness. This made the Danish people favour Sweyn over him, making it almost impossible for Harald to become King of Denmark. Realising this, Harald and Sweyn eventually sued for peace. They met at Göta river where they reluctantly signed a final peace treaty.

His harshness may have lost him popular support in Denmark, but it did not lose him battles. Harald demonstrated impressive military capabilities. He was certainly one of the finest warlords of his time. After all, he had observed and studied Greco-Roman tactics for years, and applied the knowledge to the field. He drilled his men hard, making them disciplined and tough. His army moved swiftly. He also led them with good logistical skill and generalship.

His holistic leadership can be exemplified in the battle of Vanern, where Harald defeated Earl Haakon Ivarsson. The two opposing armies were stationed each on a hilltop, with a marsh separating them. Seeing that the rebels were ill-equipped for winter, Harald understood that time was on his side, and waited to let the intense cold freeze them. After long periods of unbearable cold, the demoralised rebels struggled across the marsh to attack, but were soundly defeated by Harald's fresh and warm troops.

Norway Reformed

King Harald Hardrada was not only a skilled commander on the field, he was also an efficient administrator. His reign saw the implementation of several crucial reforms. First, he finalised the centralisation process by disbanding defiant earls and replacing them with his own, loyal nobles. This involved a final conclusion to the feud between St. Olaf and his enemies. Kalf Arnesson returned to Norway on the promise of forgiveness, but Harald tricked him. During the Danish war, Harald deliberately sent Kalf into a combat he would not survive. Only when Kalf lay slain on the field did Harald intervene with full force – a cunning play.

Having consolidated the monarchy, dealt with rebels and past foes, Harald finally had time to enact his new reforms unhindered. He developed Norway's first independent currency, helping to establish a coin-based economy in Norway. This was probably one of his most important achievements as it allowed Norwegian merchants to trade internationally much more effectively. In accordance with this, he opened up new trade routes to Russia and the Byzantine Empire, where he had a network of useful contacts to help propel the trade. Further, Harald allegedly founded Norway's current capital, Oslo, where he spent much time, possibly to monitor the trade with the Russians and have quick access to the Kattegat and Skagerrak seas.

The King made remarkable contributions within the cultural and religious sector as well. Numerous churches were built, and Harald imported Eastern Orthodox priests, teachers, monks and bishops from Russia and Greece. Harald favoured a national church policy, meaning he was suspicious of the Pope and his interventions. However, many Norwegian bishops were loyal to the Pope and were frustrated with his adherence to Orthodox Christianity. When they confronted him, Harald angrily banned them from his court and shouted that there are no other lords over Norway than the King himself. This was the beginning of a very complicated relationship between Norway and the Papacy.

Harald also spent time on the great seas. He took a gathering of huscarls and his son, Magnus Haraldsson, to the Orkneys and Shetland islands. There they subdued the earls and merged the islands into the Norwegian realm. Personally, Harald liked to sail North and North-East, searching for new and mysterious lands. According to the medieval historian, Adam of Bremen, Harald's expeditions reached the edge of the "*savage world*", where they encountered deep "*darkness*" and empty wastelands (scholars believe this was Novaya Zemlya). Adam further noted that there were rumours that Harald had also visited Vinland (America) to see the Norwegian colonies there, but such rumours were never confirmed and are unlikely.

The Great Game for England in 1066

The idea of a recreation of Cnut the Great's North Sea Empire was still very much alive. Both Sweyn and Harald thirsted for it, but out of the two, Harald was definitely the one who was most likely to get it. Norway had become the mightiest military Kingdom in the North and Harald was the most dreaded man of his time. If he were to restore the North Sea Empire, he needed to reclaim England.

He already had a convincing claim to the English throne: through his treaty with Magnus, he regarded himself as the inheritor of Magnus' claims, and Magnus had a perfectly valid claim on England's throne from his own treaty with Harthacnut in 1038 (Treaty at Göta river). Besides this, Harald held that Anglo-Saxon King Edward the Confessor, who was childless, had promised Harald that if he prevented Viking raids into England, he would become the heir to the throne. Harald had done his part, so he expected his reward.

But he was far from the only one. The Duke of Normandy, William the Bastard (who's great-great grandfather was Rollo), also had a legitimate claim. William's great-aunt was Emma of Normandy, who had first been married to King Ethelred of England and, when Ethelred died, re-married to Cnut the Great. Emma was Edward's mother via Ethelred, and Harthacnut's mother via Cnut. She was therefore the central link in this hereditary web. Since Edward the Confessor could not produce a son or daughter, he allegedly promised to make William his heir.

However, when Edward the Confessor passed away in January 1066, neither William nor Harald were declared successors. Instead, it went to Harold Godwinson, son of the influential Earl Godwin. This was an outrage – Harold Godwinson was not of royal blood and had no hereditary legitimacy. Both King Harald and Duke William now mobilised their forces for war.

It was time for Harald's war machine to face its ultimate test. In Norway, he assembled around 240 warships, which would count for some 10 000 soldiers. Before leaving, he declared his firstborn son Magnus Haraldsson as King of Norway while he was away. He then embarked the ship *Long Serpent* together with his Queen Elizaveta, his daughters and his second son, Olaf the Elegant.

They first sailed through the Orkneys and Shetland islands where the North Sea Earls joined him with 2000 men. While passing by Scotland, King Malcolm III gave him an additional 2000 Scotsmen. When they finally disembarked on English soil, they numbered some 17 000 men – the largest army any Norwegian had assembled at that point[27].

Harald was also aided by the Anglo-Saxon Earl Tostig Godwinson. Tostig was Harold Godwinson's brother, but his disorderly leadership of Northumbria had alienated him from the family and turned him into an outlaw. Following this, Tostig allegedly visited both King Harald and Duke William to seek support. Some sources claim that he was the one who persuaded Harald to make the invasion, as Harald was initially uninterested. The exact sequence of events remain unclear, but whatever the case, England now faced a two-front invasion. King Harald was invading from the North-East, and Duke William from the South. Harold Godwinson found himself in a grim situation.

In September 1066, Harald landed on the shores of Northumbria. He attacked a nearby fief, and when the fief showed resistance, he burned it all to the ground, sending a clear message to all who planned to oppose him. The tactic worked. All nearby fiefs quickly surrendered.

Harald initially wanted to continue sailing by river towards York, but the river became narrower closer to York, causing impracticalities for Harald's huge fleet. He decided to set up a defensible camp near the Riccall, close to the Fulford-road that would take him to York by foot.

Two Anglo-Saxon Earls, Morcar and Edwin, now mobilized their respective armies for battle. As King Harold Godwinson was busy mustering his own soldiers in the South, they hoped to win glory and respect by stopping the invaders themselves. They marched at Fulford Gate where they invited the Norwegians for battle. Their invite was accepted.

---

[27] Modern historians speculate that in the 11th century it would have been impossible to supply such a large army with food, and so argue that his armada was significantly smaller. This is an ongoing debate.

Harald concentrated his trained warriors on the centre and left flank, deliberately exposing Tostig and his men on the right flank in a thin, long line. The strategy was to lure the enemy to attack Harald's right, drawing them out of formation, tiring them, and then for Harald to lead his disciplined troops from the centre and left in a relentless charge. Once the battle began, everything went according to plan. Tostig's line held and Harald's troops stormed at the enemy centre, breaking their lines, and sending the entire army in an all-out rout. It was a triumph.

Of course, Harald could let his men loose to pursue the enemy into York and sack it – but this would be reckless. York was home to a vast Norse population that could possibly provide him with manpower. Turning them against him by plundering would be very unwise. Besides, he intended to use York as his headquarters for further campaigns and had promised Tostig the Earldom of Northumbria – he therefore needed to annex the city in the most peaceful and orderly manner possible.

Thanks to the discipline of his army, he rallied them under strict command. Then, his entire army paraded in front of York and demanded its surrender. The noblemen of the city were dressed in their finest robes as they came to witness the spectacle. There was nothing they could do. They opened the gates and surrendered the city. York was his.

One final brick had to be put in place for all of Northumbria to be secured. Harald needed to ensure the allegiance of the nobles. He therefore arranged with them that he would take 150 of their sons as hostages, while they would take hostage 150 of Harald's men in return. This swap was agreed to take place at Stamford Bridge.

What Harald was completely unaware of, was that King Harold Godwinson had, to the amazement of modern scholars, marched *80km North a day* since Harald first arrived. His entire Saxon army was encamped just 16km away from York. It was an unfathomable achievement.

Battle of Stamford Bridge

On the morning of September 25th, 1066, Harald and Tostig left camp at Riccall and rode towards Stamford Bridge. Throughout his career, Harald had always taken serious precaution ahead of any mission – but on that day, he seemed overconfident. He refused to use his agents to scout ahead. He brought with him only a portion of his total army (ca. 10 000 men) – the rest (ca. 6 000) probably still sleeping in their tents. Additionally, thinking they were only going to a harmless meeting, none of the men wore armour.

Harald and Tostig arrived at Stamford Bridge and awaited the nobles and their proposed hostages. Shortly after, they witnessed something totally unsuspecting: the main 15 000-strong army of King Godwinson. Some came fleeing from the woods, shouting the news of the Saxon arrival. This handful of Norwegians then took up position at the bridge, sacrificing themselves in order to give King Harald enough time to prepare for battle.

One of these brave men was a massive berserker. After a few moments of fighting, he was the last man protecting the bridgehead, and that for a reason: he seemed impossible to kill. He swung his double axe and slaughtered scores of intimidated Anglo-Saxons. During the small pauses in-between, the Saxons tried to appease him by offering bribes, but he flatly rejected and mocked them. They shot an arrow through him, but he broke it off and continued fighting. Two Saxons then snuck under the bridge and pierced a lance through him from underneath. He finally collapsed and the Saxons stormed over the bridge. It must have made a profound impression on those who witnessed it.

In the meantime, Harald had quickly readied his men and taken up position on a hill, ordering them to form a circle formation. Tostig proposed a hasty retreat to the main camp at Riccall, but Harald denied it. He knew that the Saxon cavalry would cut them to pieces along the way. Besides, King Godwinson could easily close the bridge at Kexby, blocking the road to Riccall. Instead, Harald had sent three couriers on horseback to call for aid. His only hope was that he could hold off the Saxons until the rest of the army arrived.

The Saxon heavy cavalry opened the main phase of the battle by smashing into the Norwegian shield-wall. They tried consistently to break the line but failed. The shield-wall held, and the Norwegian archers let arrows rain down on the enemy. There were huge losses on both sides.

Harald then changed his strategy. Believing that his shield-wall would be unable to hold until reinforcements arrived, he began leading fast, decisive counterattacks on the Saxon cavalry to scare their inexperienced conscripts away. With no armour, the King personally led these attacks, throwing himself amid lethal danger. He cut down his enemies like a berserker - but suddenly an arrow pierced his throat. The blood was everywhere as Harald fell to the ground and gasped for air. As he lay motionless on the ground, King Harald Hardrada drew his last breath. Seeing this, Tostig lost his nerve. He immediately sounded retreat – causing much disarray as the formation disintegrated.

A group of steadfast huscarls formed a last stand around the lifeless body of their dead King. Harold Godwinson, who must have admired the valour of these men, offered them peace and quarters if they surrendered – but they shouted back that they would rather die than surrender. Arnor, a Skald, sung: *"The King, whose name would ill-doers scare, the gold-tipped arrow would not spare. Unhelmed, unarmoured, without shield, he fell among us in the field. The gallant men who saw him fall, would take no quarter, one and all. Resolved to die with their loved King, around his corpse in a corpse-ring."* The Saxons charged at them and fought them for hours.

A group of elite warriors, led by Harald's marshal Eystein Orre, now arrived at the field of battle, only to see the valiant last-stand and the body of their beloved King. Enraged, Eystein attacked the exhausted Saxons. The battle continued, and the fatigued Saxons showed signs of faulty. However, Eystein failed to deliver a decisive blow. At nightfall, Eystein was killed along with the huscarls and the rest of his men fled. The battle ended.

Harold Godwinson reached the main encampment at Riccall but didn't attack. Instead, he spoke to Olaf the Elegant, Harald Hardrada's son, and discussed a peace treaty. The Norwegian captives were released, but in exchange, Olaf would promise to leave immediately and never return to avenge his father. It is likely he was also forced to give up Hardrada's famous Byzantine gold [28]. The young, 17-year-old Olaf accepted these terms.

The Norwegians had suffered around 6000 dead, including many of their nobles and their own King. It was a bitter defeat. Olaf picked up the body of his deceased father on the bloody field near Stamford Bridge. He then dismissed the Scottish and North Sea-mercenaries and took the remaining Norwegian warriors back to Norway. The battle was one of the hardest fought in Anglo-Saxon history – a pure bloodbath. The dead bodies lay unburied. Their bones still scattered the hill years after 1066.

Harold Godwinson was not able to repeat his hard-won success. When he arrived at Hastings to face Duke William, his army was exhausted, battered, and rugged from the unforgiving fight against the Norwegians. Harold Godwinson lost and Duke William, now called William *the Conqueror*, won the throne of England. The Normans have ruled England ever since.

---

[28] Hardrada's Byzantine gold had formed the basis of his finances for decades. Even by 1066 it was extraordinarily valuable. The gold was captured by William the Conqueror and many, like Adam of Bremen, argue it served as William's main financial source in the opening years of his reign. William bribed many for peace, securing his foothold on England.

Legacy

Harald the Wolf-feeder, the Bulgar-burner, the Land-waster, the Harsh-ruler. Harald Hardrada was one of Norway's exceptional characters. His adventurous life covered the drama at Stiklestad, the parched sand dunes of Arabia, the revolutionary night in Constantinople, and the brutal wars in Scandinavia, to the valiant combat at Stamford Bridge. His name is traced in Norse, Greek and English writings, as he affected the course of history in all these theatres. The Battle of Stamford Bridge, in particular, has been immortalized as a crucial element of the 1066-spectacle that helped William the Conqueror win at Hastings. Consequently, a new chapter in world history began: the rise of Norman England.

Harald Hardrada was also the monarch who ushered Norway into its first golden age. He conducted vital reforms – like the introduction of a national currency – founded Oslo, and established lasting trade routes over the continent that boosted the country's economy. His expansion and improvement of the military was an equally astonishing feat.

Perhaps more significantly, Harald Hardrada ended the mighty influence of the earls once and for all. The dynasty of the Earls of Lade had ended. In this way, he completed the work of his brother Olaf Haraldson, turning Norway finally into a feudal, European-like country. This would give future Kings of Norway increased stability, structure and security.

However, this socio-cultural revolution came at a heavy price. Both Olaf and Harald stirred enormous controversy as they force the Norwegians to accommodate the new state system and religion. Harald Hardrada, in particular, was infamous for his harshness. Though this made him feared by his enemies, it also complicated his relationship with the people.

Perhaps the best description of these complicated characters is from Haldor Brynjolfson, a chief who was a close friend of both brothers. He described them in the following way: *"Both* [Harald Hardrada and Olaf Haraldsson] *were of the highest understanding, and bold in arms, and greedy in power and property; of great courage, but not acquainted with the way of winning the favour of the people; zealous in governing and severe in revenge...Both brothers, in daily life, were of a worthy and considerable manner of living; they were of great experience, and very hard-working, and were known and celebrated far and wide for these qualities."*

King Haakon the Good at the *blot* on Maere. Painting by Peter Nicolai Arbo.

Olaf Tryggvason is proclaimed King. Painting by Peter Nicolai Arbo.

St. Olaf falls at Stiklestad. Painting by Peter Nicolai Arbo.

Harald Hardrada's last moment at Stamford bridge. Painting by Peter Nicolai Arbo.

Sverre crosses the mountains of Sogn during winter. Painting by Peter Nicolai Arbo.

The Birchlegs save Haakon Haakonsson. Painting by Knud Bergslien.

Orthodox icon of Saint Olaf. Painted by the hand of Vappu Larsen.

PART II

Olaf the Peaceful & Magnus Barefoot

1050 – 1093 & 1073 – 1103

Norway was ready to begin its next chapter. The country was unified, ready for the Kings and noblemen to steer it onwards. These rulers needed to be wise and righteous, for they now held enormous responsibility. In many ways, Olaf the Peaceful set an important precedent as a good King. If future rulers could be as honourable, kind-hearted, and wise as he, then surely, Norway would rise to become a jewel in the North.

Meanwhile, his son, Magnus Barefoot, would represent the return of Viking aggression. His ambitions were equal to those of his grandfather, Hardrada: he desired to live like a true warrior, sailing on the vast oceans and conquering new lands. His reign bolstered the idea of Norwegian expansion – attempts to subdue all lands of Norwegian heritage under one Crown.

Olaf *Kyrre* and the peaceful years

Witnessing the aftermath of the Stamford bridge carnage made a lasting impression on Olaf Haraldsson. He helped carry his dead father back to Riccall, where the camp was seized by the victorious Anglo-Saxons. As a Prince of Norway, Olaf assumed responsibility to broker an agreement with Harold Godwinson. They were allowed to leave but had to swear never to return and probably pay out a ransom. Having dealt with the English, Olaf took his mourning family and remaining Norwegian army and headed for Norway. His sister, Maria, did not survive the winter and died later that year. The sagas claim she died naturally upon learning of her father's death (perhaps she was already dying). Queen Elizaveta, Harald's wife, also died the following year in 1067.

All of this made Olaf despise war. He probably saw nothing glorious about it, only death, misery, and destruction in search of some vainglorious goal. When he returned to Norway, he immediately initiated a peace process. He and his older brother, Magnus Haraldsson, cancelled their inherited claims on Denmark and instead formed a peaceful alliance with Sweyn Estridsson. The peace was sealed with Olaf's marriage to Sweyn's daughter, Ingerid of Denmark.

Shortly after, Magnus Haraldsson fell ill of disease and died, leaving Olaf to assume solitary kingship of the country. Olaf would go on to safeguard the peace by consistently rejecting war in favour of trade. For example, when King Cnut II of Denmark invited him to an invasion of Norman-England, he again declined. Instead, he established trade routes with England that subsequently led to enhanced prosperity for both countries.

Olaf Haraldsson *"Kyrre"*, meaning *"the Peaceful"*, led a consolidation-policy. He reduced the amounts of provincial laws to centralise the legal system. He founded myriads of new merchant towns along the coast, with the largest one being Bergen, the later capital of Norway.

Olaf was an active church-builder. Under his watch, the famous *Nidarosdomen* (Cathedral of Nidaros) was founded to honour St. Olaf. In Bergen, *Kristkirken* (Christ's Church) was constructed. In addition to these, Olaf erected numerous churches across the country and facilitated the construction of monasteries. His reign saw a reconciliation with the Church, granting it more autonomy and privileges. Pope Gregory VII wrote letters to Olaf Kyrre, symbolising the inclusion of Norway into the world of Christendom.

Olaf Kyrre facilitated the establishment of a national monetary institution, based on counting instead of weighing coins. This made the economy easier to manage and monitor. Amusement shows were often arranged for public entertainment, and foreigners from across Europe began visiting the trade cities of Bergen and Nidaros.

Furthermore, Olaf built guilds across the country. These guilds acted like town halls and enhanced provincial autonomy. Guildsmen raised finances and invested in their communities. Wealthy merchants also met in the guilds to feast together, discuss business, and venerate the Saints. As it was illegal to drink alcohol in public, many businessmen used the guilds as drinking places.

King Olaf himself often came to join them and drink. One day, when they noticed how merry he was, he held a speech for them, saying: *"I have reason to be glad when I see my subjects sitting happy and free in a guild consecrated to my uncle, the Saint King Olaf. In the days of my father, these people were subjected to much terror and fear. The most of them concealed their gold and their precious things, but now I see glittering on his person what each one owns. Your freedom is my gladness."*

Olaf's strong fashion sense was amplified when he assumed power. He decorated his courtroom with precious ornaments and exotic materials, promoted a butler, torch bearers, and arranged a staff of 60 servants for the royal residence. The court consisted of 120 men and 60 pursuivants, all of which were expected to be appropriately dressed in fine coats, ties, and shiny boots. Olaf had always had this sense of fashion – in his youth, he had been known as Olaf the Elegant.

> *"Our Trondheim King is brave and wise, his love of peace – our people's prize. By friendly word and ready hand, he holds good peace through every land. He is for all a lucky star, England he frightens from war. The stiff-necked Danes he drives to peace, troubles by his good influence they cease,"*

– sung the poet Stein.

Olaf Kyrre the Peaceful certainly left a loving legacy. The country was in dire need to recover from the perpetual unrest and wars of the past century. In this way, his reign healed Norway's wounds.

It is likely to also have led to a paradigm shift in the way Norwegians viewed the new monarchical state structure and the legacy of St. Olaf. Olaf Kyrre's reign demonstrated how efficient this feudal system could be in generating wealth and societal harmony. Wise and responsible kingship could advance the country to new heights and increase the welfare of the people. The sagas say that he improved the livelihood of thousands of Norwegians, and generously gave away riches and fine clothing to his colleagues and servants.

> *"He was a wise man, and well understood what was of advantage to the Kingdom...his reign there had no strife, and he protected himself and his realm against enemies abroad. His nearest neighbours stood in great awe of him, although he was a most gentle man...Norway was much improved in riches and cultivation during his reign,"*

– writes the saga-writer Snorri Sturlason.

If the unified Kingdom led to such success in peacetime, then what could it achieve in wartime? This was the question Magnus Olafsson, King Olaf's bastard son, must have pondered. He dreamed of plunging into open seas and conquering foreign lands like his hero grandfather Harald Hardrada. The young prince had many ideas: he envisioned an expanded, Norwegian realm – a North Sea empire.

Magnus Olafsson's path to the throne

In 1093, prince Magnus Olafsson sailed the North Sea, patrolling its islands and inspecting their conditions. One day he received the breaking news: King Olaf the Peaceful was dead. Magnus rushed home. Though he was the official heir, he was a bastard son, which, according to church law, denied him the throne.

Magnus ignored these customs, but many aristocrats did not. Haakon Magnusson, Magnus' cousin, laid claims to the throne. He was firmly backed by aristocrats in Trondelag. Magnus assembled an army and marched North. Civilians now dreaded a relapse into civil war – but fortunately, the cousins avoided war by agreeing to co-rule instead.

Just two years later, in 1095, Haakon died of disease – but Haakon's supporters refused to accept Magnus as solitary King. Clearly, they could be considered rebels, so Magnus treated them as such. He outmanoeuvred them in a quick conflict, captured them and hung them all. It is said that Magnus was saddened by the sight of the dangling corpses. These were good men, he held, and it was a shame that their lives had to be wasted like this, as rebels. Nevertheless, by 1095 the throne of Norway was fully in Magnus' hands.

His first acts as sole King were administrative. He adjusted the monetary policy (coins were to be produced with 90% silver, not 30% as Hardrada had started with) to strengthen the Norwegian currency, strengthened the network of the Church and imported new priests and bishops from Germany. He also solidified the feudal system by only promoting close family members as aristocrats and nobles. However, Magnus would not spend most of his life by a desk doing paperwork – he was destined out in the open seas, seeking conquest.

Magnus the Conqueror – first expedition West

Ever since the dawn of the Viking Age, Norwegians had populated the North Sea isles, carved out their own kingdoms, earldoms and chiefdoms and made their own laws and customs. In their days of glory, they were prosperous, but by the end of the 11th century, their decline and fall seemed inevitable. Norse rulers tore their lands apart with civil strife and blood feuds. Viking and raiding bands often harboured the islands, disturbing and impeding trade. Some Norse communities were at the verge of extinction, as they no longer had the capacity to survive as cohesive societies. The plight was serious.

It was time for these long-lost brethren of Britannia to reunite with the natives of Norway: for all Norwegians to unite under one banner. It would facilitate greater trade, bring in substantial tax returns, and elevate Norway's status in Britannia. It was a grand vision – but could he realise it?

In 1097, Magnus mustered 100 warships and sailed across the North Sea. It was the first major military operation overseas since 1066. His first target was the independent Earldom of the Orkneys, which had long exerted great power over the region, but now was just a shadow of its former glory. Magnus easily toppled its Earls and installed his own. He then assaulted the Hebrides, ravaging various islands to quell any thoughts of resistance. Many chiefs surrendered without a fight. The only island Magnus did not threaten was Iona. This was the resting place of St. Columba, a holy man whose monastery had been wrecked by Norse Vikings in 794. Magnus paid homage by his tomb, and then distributed gifts to the chiefs of the island and won their loyalty. Iona was subdued without a single drop of blood.

He marauded his way down South, seizing Irish and Scottish fiefs along the coasts. The next target was the Isle of Man – an island situated in the heart of the Irish sea, making it ideal as a springboard for future operations. Magnus invaded, but faced stubborn resistance by its ruler, Earl Ottar, a former friend of Harald Hardrada. It was a tough fight, but the Norwegians prevailed in the end. Magnus sent word home, encouraging mainland Norwegians to come settle on Man, and built a major fortress there to use as his headquarters.

Maintaining momentum, Magnus continued further South. His ships silently patrolled the coasts of the island Anglesey – the sixth largest island of Britannia, situated in Northern Wales. One day, his scouts spotted a large Norman army led by Hugh the Stout. Magnus immediately readied his men and charged their camp. The Normans naturally did not expect such a shock attack, and were completely overrun. Magnus himself shot Hugh with his bow and the Normans routed. Anglesey was then absorbed into Magnus' ever-expanding realm.

This shock attack on the Normans was not simply a reckless act - it was part of a wider, geopolitical goal. Magnus had realised that the Norse communities around the Irish Sea were hesitant in joining him, fearful of retaliation from the Normans. Since 1066, the Normans dominated Britannia. Rebellions against them were easily crushed, and their military outmatched any other. In the 1090s, they pushed into Wales, but were stopped by defiant Welsh patriots. In 1098, King Griffith of Wales had visited Magnus, urging him to strike the Normans and help him drive them away. In return, he could keep conquered territory and win favour with Wales. Magnus knew that a victory over the Normans could also sway the local Norsemen to rally to him. His successful conquest of Anglesey therefore achieved all this.

King Magnus now had a widespread reputation. He had thundered through Britannia with unstoppable force. As he began sailing North again, King Edgar of Scotland knew he had to come to terms with this sudden, new player in Britain. In the negotiations, Magnus held strong leverage. Edgar was weakened by internal revolts, and he suspected that Magnus had connections among Scotland's nobles. In the end, Edgar agreed to cede all Scottish islands to Norway in return for peace. With this great bargain, Magnus concluded his first campaign.

The Three Kings

The King returned home in 1099 as confident as ever. He believed he had the right attributes and military to undertake great campaigns. He decided it was time to restore the old borders between Norway and Sweden, meaning the Göta river. He mobilised his forces and invaded.

Soon, King Inge of Swealand came to stop him with a superior army. Magnus' marshals suggested retreat, but the King disagreed. Once a campaign had started, it ought to be concluded. No retreat. Instead, full attack. His army marched at all haste to locate the Swedish forces. Once the scouts spotted it by Fuxerna, the Norwegian huscarls lurked around the forests and took up a strategic position. Suddenly, they shouted their warcries and began the ambush. Magnus allegedly rode fiercely into the battle, cutting down soldiers from side to side and often finding himself completely alone. The Swedes routed. King Inge escaped by the grit of his teeth. The battle of Fuxerna was a classic case of Norse warfare: when encountering a superior force, they didn't consider retreat, but instead used deception and stealth to attack the enemy when he was unfairly unprepared.

Encouraged by this success, Magnus proceeded further North, occupying Vastergotland, which he argued traditionally belonged to Norway. By the end of the year, Magnus let build a fortress at Kallandsø and deployed 300 warriors to guard it for the winter. He himself returned to Norway, and promised to resume the campaign in spring.

However, with Magnus gone, King Inge swiftly retaliated. With 3000 men he conquered the fortress at Kannadsø and humiliated the Norwegian garrison, sending them limping back to Norway. Infuriated over this embarrassment, Magnus relaunched his offensive. Combats were tough, and Magnus was reckless. He was well recognisable on the field and therefore hunted by his enemies. During one collision in particular, the Swedes trapped and almost killed him. He slipped away from certain death only when his loyal friend Ogmund confused the Swedes by pretending to be the King.

Eventually the war spun out of control. Nobles on each side tried to appease their Kings, but with no success. In the meantime, Danish King Eric worried that the war would escalate into Danish territory. The violence had to stop. Eric gathered Magnus and Inge to Göta river to negotiate peace.

This became the famous meeting of *the Three Kings*. The peace treaty between them became very important for Scandinavia because it finally defined and acknowledged the borders and territories of each sovereign country. To seal the new relationship, Magnus married King Inge's daughter.

Quest for Ireland

Meanwhile, Magnus heard news that his governor of Man had been killed by rebels. His holdings in Britannia were at risk, and it was clear that he had to return to enforce discipline. He assembled an even greater army than before and set sail. Some may have believed he did all this to deal with some rebels – but Magnus merely used this as a convenient entry back into Britannia. His real intensions were much more ambitious: he planned to make himself the High King of Ireland, the final piece in his vision.

Norwegians had a historical heritage on Ireland. They had founded Dublin and ruled it as an independent Norse Kingdom for centuries. Their golden years were in the early 900s, when they expanded their holdings in all directions, even capturing York. However, the sleeping giants of Ireland slowly awoke. In 1014, Irish King Brian Boru forged an Irish coalition against the Norse Vikings of Dublin. At the bloody battle of Clontarf, the Irish won a costly victory over the Vikings, marking the decline of Norse power in Ireland. They began paying tribute to Irish Kings of Munster, took local wives and adopted Irish customs and traditions. Then, in 1102, Magnus came sailing in. Magnus resonated with this heritage by adopting the dressing style of Irish warriors who traditionally fought barelegged, earning him the odd epithet *Barefoot*.

However, though Magnus and his men were famed for their toughness, Ireland would be a difficult land to subdue. The island was large, and the Irish were plenty in number and courageous at heart. One of Magnus' subordinates voiced his concern, referring to previous Norwegians who had attempted the same, but failed. Magnus allegedly countered his words by affirming: *"Work started must always be finished. Bravery and great effort will yield progress."*

The first stage of the Irish conquest opened when Magnus demanded the current High King, Muircheatach, to submit to Norway. He sent his shoes to Muircheatach, telling him to gather his subordinates and place the shoes on his shoulders to symbolize his surrender. According to the sources, without protest, Muircheatach did exactly so. Magnus had his son, Sigurd, marry Muircheatach's daughter. This marriage won him the major city of Dublin and already made him, alongside Muircheatach, the most powerful man on Ireland.

Operating from Dublin, Magnus led several campaigns in the area, especially against Donald O'Lochlain, the arch-enemy of Muircheatach. A key element of the agreement with Muircheatach and Magnus was for the latter to help defeat Donald. While the Norwegians did the dirty work on the battlefield, Muircheatach promised to provide supplies.

However, there was an eerie feeling that trouble was aloof. So far, it had been *too* easy. Was Muircheatach sincere, or was Magnus being tricked?

While Magnus campaigned against Donald, the provision of supplies mysteriously stopped. Magnus took a detour to enter Muircheatach's lands and investigate the matter. Suddenly – on the 24th of August, 1103 – the Norwegians were ambushed by a hailing swarm of Irish warriors, hurling at them and hacking them to pieces. The Norwegians were stuck in difficult, swampy terrain. Arrows and spears flew in the air as Magnus took control over the shocked troops and organized them into a shield-wall.

His companion Eywind advised the King to leave the men and flee to the ships, but the King replied: *"It is not honourable of a King to abandon his men, these good men who follow me. Let us instead unite to face the Irish!"* Magnus scanned the field of battle and figured a way out. He noticed a ridge a short distance away and ordered his subordinate Thorgrim to split with the army with a group of soldiers and occupy the ridge. Magnus would then slowly make his way to the ridge, where he could then continue the battle in a better position.

But Thorgrim waivered. He retreated before Magnus could reach the ridge. Enraged, the King screamed after Thorgrim: *"You are deserting your King in an unmanly way! I was foolish in making you huscarl, and driving Sigurd the Hound [29] out of the country, for he would have never betrayed me!"* The situation was now critical as the Norwegian position was very fragile.

---

[29] Sigurd Hund was an aristocrat Magnus outlawed for treason.

They fought valiantly but hopes for survival diminished every minute. Suddenly a spear pierced through both of Magnus' thighs. He held back the immense pain, pulled the spear out of his own legs broke it, and continued to direct his troops. Realizing the uneasiness of his men as they saw his mortal wounds, he shouted: "*thus we break these twigs, my lads; let us go briskly on. Nothing hurts!*" Shortly after, an Irish berserker lynched at Magnus and chopped him in the neck with an axe. Magnus Barefoot died instantly. He was 30 years old. Leaderless and demoralized, the rest of the army fled to the ships.

The funeral took place on the following day. Attending it were the Norwegian huscarls who had survived the battle, numerous Irish chiefs and even Muircheatach, who wished to witness the burial of the Norwegian thunderbolt. Magnus' companion Vidkunn surrendered the royal sword to Muircheatach, dramatically symbolising the end to the Viking age. This was the will of King Magnus. If he were to die, Muircheatach was to inherit his sword. However, behind the formalities, Vidkunn and the other Norwegians must have understood what had happened. Muircheatach engineered the ambush in Ulster that had killed their King.

The 24th of August 1103 was indeed the final ending moment of the Norwegian influence over Ireland. Dublin fell to the hands of the Irish. The Vikings would never return.

Legacy

> "*Til frægdar skal konung hafa, en ekki til langlifis!*" (in Norse)
> "*The Kings are made for honour, not for long life.*"
>
> – Magnus Barefoot's motto

This motto would become Magnus' self-fulfilling prophecy. It was an inevitable end to the young King, and perhaps the type of end he wanted, as he always threw himself into combat and great danger. Magnus wanted to awaken and maintain the Norse warrior culture, where the chief was always expected to lead the charge and where victory was secured through ferociousness and trickery. However, this time, the end goal was not valuable loot – it was to expand Norway's borders to far-away places and create a new, North Sea realm.

His lightening offensive into Britannia took the region by surprise, and it firmly established Norwegian sovereignty in Britain. The claims he secured were important in defining what territories could be considered *Norwegian*. The Treaty of the Three Kings also secured and acknowledged Norwegian authority all the way to Göta river. This had lasting impact in the way the Scandinavian countries regarded each other.

For 12 consecutive years after the ambush in the Irish swamp, the Norwegian Royal court received gifts and honours from Ireland to commemorate the fallen Magnus Barefoot. The Scots wrote ballads about the King, praising his exploits and deeds. In Irish poetry, *Manos Mor* (his name in Gaelic), is mentioned frequently, as he is an exciting figure in Irish history. This was the King who shared his soldiers' hardships, put himself in equal danger and suffered the same wounds.

On the day of his death, Magnus wore a red armour with a very special icon woven to his chest. It was a golden, crowned lion holding an axe. This was the symbol of Norway's *Eternal King*, St. Olaf. In the context of Magnus' death, this symbol was immortalized. Today, Magnus Barefoot's shield is Norway's official Coat of Arms.

Sigurd the Crusader

1090 – 1130

Sigurd the Crusader was a new phenomenon in European history. For the first time, a European King personally led a crusade. This made him highly respected, but also signalled to the rest of the continent that Norway was a robust and reliable Christian Kingdom that intended to serve the will of the Papacy and Europe. He embarked on an adventure that would change both him, and Norway, forever.

Europe's First Crusader King

Sigurd was a son of Magnus Barefoot. Since he had shown interest for the military at a young age, Magnus sought to give him the full experience of it. He brought him on his campaigns in Britannia, where little Sigurd probably witnessed distressing violence at a tender age. The little boy also became a pawn in his father's geopolitics. He was installed as the nominal Earl of the Orkneys and Shetlands at age eight, and, at age 13, married off to the daughter of Muircheatach for political reasons. Upon his father's sudden death, Sigurd was only 15. He obviously fled from Ireland, reaching Norway in 1103. That year, he officially became King together with his two brothers, Eystein and Olaf. Olaf was just an infant, making Eystein and Sigurd the effective rulers.

Sigurd had by then witnessed more of war and treacherous politics than the average adult. This must have weighed heavily on his mind, and he probably sought relief in Christianity. Sigurd was a devout Christian and placed much attention to his faith. He had a burning desire to visit and pray by the many holy sites of this world. It is unknown how much theology he knew, but he certainly showed an eager to serve Christendom the best he could. However, Sigurd was no priest – he was a warrior. Waging war was what he knew, and at the time, it was all he needed, for this was the age of the Crusades.

The crusader project originally began when Pope Urban II called for the Europeans to mobilise and attack the Muslim holdings in the Middle East. Ever since the 7th century, Muslim armies had invaded, occupied, and plundered many parts of the Christian world – first the regions of Palestine, Syria, Egypt, Libya, Carthage and the rest of North Africa, but then also Spain, France and Italy. Their invasion was stopped by the

Franks at the battle of Tours/Poitiers in 732, but the wars continued. In 846, they sacked Rome, looting the Old St. Peter's basilica and other churches. By 1090, the Byzantines were struggling to keep the Muslim Turks at bay, and the Greek Emperor pleaded his European brothers to come to his aid. A united European offensive could drive the Turks away and reclaim the lost lands of Christendom, especially the so-called *Holy Land* (modern-day Israel, Jordan, Lebanon and Syria). Besides, Muslim authorities had disallowed Christian pilgrims from entering Jerusalem. Pilgrimage was a sacred right and had to be defended, it was argued.

The First Crusade (1095 to 1099) was a success that led to the establishment the Kingdom of Jerusalem. However, after its founding, most of the warrior-pilgrims returned to Europe, leaving the Kingdom short on manpower. A plea was sent for all Europeans to become soldiers of Jerusalem and protect the nascent state. Who would have known that this plea was to be answered by the most distant of them all: Sigurd of Norway.

The young King, brought up with the calamities of war, fervently prepared a crusader army to embark on his exotic journey south. Sigurd mustered around 1500 men and set sail in 1108. In his absence, his elder brother Eystein would govern Norway. Sigurd became the first crusader King in Europe – a startling feat. Who would have imagined that the pagan Viking people would one day be at the forefront of Crusader activity? Keep in mind, this was just 78 years after St. Olaf's death at Stiklestad.

The Long Journey

The zeal of a crusade attracted many commoners and soldiers to Sigurd. As he paraded through Europe, his force may have grown to around 6000 men. Additionally, the Kings of Europe gave him military access, funds, horses, food, other supplies and, sometimes, company. Sigurd attended banquets with both the King of England, Henry I, and the King of France.

In Galicia, he made a personal visit to *Santiago de Compostella* – the holy grave of St. Jacob, one of Christ's Twelve Apostles. Ironically, the Lord of Santiago de Compostella refused to supply Sigurd's army with food, so Sigurd had to attack the castle and loot it for supplies. When he returned to the sea, another threat appeared on the horizon: galleys of pirates coming to ambush him. Only after fighting them off could the crusaders continue further South to Portugal. These initial skirmishes served as a dire warning: they were now entering hostile, Muslim-held territories.

The Iberian peninsula was at the time divided between Catholic Spanish kingdoms and Muslim Moors and Berbers. Tensions were especially high in Lisbon, a city divided between Christians and Muslims. To weaken the latter, Sigurd launched an attack on the

Muslim part of the city, winning a major battle and driving them away. He then sailed Southwards, patrolling and capturing several forts along the way. Seville was besieged, but abandoned after the siege began draining his resources. Instead, they passed through the Strait of Gibraltar, heading towards the islands of Formentera, Ibiza and Menorca.

Here, Sigurd fought three consecutive battles against African pirates, one of which was won using very unorthodox tactics. After a skirmish, the pirates retreated into a cave in a cliff. Sigurd and his crusaders tried numerous times to enter the cave, but many fell off mountain wall. So, they instead filled two longships with soldiers and lowered them to the mouth of the cave with thick ropes. The pirates retreated deeper into the cave, but Sigurd, sensing that this was a trap, held his troops back. Instead of pursuing them, he ordered nearby trees to be cut down to start a large bonfire by the mouth of the cave. The fumes turned the cave into an inferno. Some pirates choked by the smoke, others came running out only to be killed by the crusaders. This happened in Formentera, and the cave is known as the *Cave of fire* to this day.

The voyage continued to Sicily, where Sigurd spent seven full days in the company of the Norman King Roger II of Sicily. The Normans had come to Southern Italy to drive out the Muslims and protect Rome from Arab threats. By 1090, the Normans had conquered all of Southern Italy and Sicily, and became its rulers. King Roger II of Sicily descended from Hjallt, one of Rollo's close confidants, and probably regarded Sigurd and his Norse crusaders as his kinfolk.

In 1110, Sigurd finally arrived at port Jaffa in the Holy Land. Here they were received by King Baldwin of Jerusalem. According to the medieval chronicler Fulcher of Chartres, Sigurd arrived with 10 000 men to reinforce the Kingdom.

The Holy Land

For Baldwin, news of Sigurd's arrival could not have been better timed. He was in the middle of the siege of Sidon, one of the most important coastal cities in the Holy Land. It was occupied by Arab raiders who constantly threatened Christian pilgrims and regional trade routes. Baldwin had personally left the siege to greet Sigurd. He wanted to give the Norwegian King a warm welcome and had arranged a grand tour of the Holy Land.

They rode across the historic lands until they finally rested their eyes on the spectacular Jerusalem, easily identified by the iconic sight of the al-Aqsa mosque overlooking the Jewish Wailing Wall. Just beside these places were the Church of the Holy Sepulchre – the site where Christ himself was crucified. It was a city unlike any other. Some called it the very centre of the world, the bridge between East and West,

North and South. Here came merchants from far-away places – Nubians from Africa, nomads from the Eurasian plains – and even Scandinavians from Finnmark, selling their hides and ivory and entertaining onlookers with clever wordplay. The bustling markets offered precious ornaments and jewellry; make-up for women, fine dresses, robes and shirts; swords and souvenirs; tea, ice, endless selections of spices and nuts, and hummus; medicines, oils and herbs. Suddenly, one noticed religious Jews hurrying to reach the Wailing Wall for their Sabbath, or a square filled with bowing Muslims, praying to *Allah* in the direction of Mecca. The church bells were ringing, gathering Christians from all the corners of the known world to come and kneel by Christ's empty tomb.

It must have been quite a spectacular experience for the Norwegian and North Sea crusaders. It was, of course, far from the first time Northmen had seen these cities. Varangians had been here many times. Even Norway's former King, Harald Hardrada, patrolled these lands. Norse traders had also done business in the Holy Land and the East. But it was a long journey to Jerusalem, and just the lucky few had the privilege of beholding the city. So, for the Northmen who had only seen rocky archipelagos and snowy icescapes, Jerusalem was an unforgettable sight.

Sigurd and Baldwin entered the Church of the Holy Sepulchre, where Sigurd prayed. They then rode out of the city and entered the mountainous terrain of Judea and Samaria. Here, the sands whispered stories of ancient glories and Spiritual revelations. This was the land the warrior Joshua took to be his. Baldwin and Sigurd went on to visit the river Jordan, where Jesus Christ himself had been baptised. Sigurd took a long bath here, contemplating his life and praying for mercy.

By December that year, Baldwin received reports that his soldiers had finished preparations for an assault on Sidon. Sigurd was to join forces with the Venetian fleet and attack Sidon by sea. The attack went precisely as planned and Sidon was taken. Shortly after, Beirut fell to the crusaders. By Christmas therefore, the coastline was secured.

Sigurd now decided to leave the Holy Land. He could have stayed and made quite the career as a crusader commander, but he had his own Kingdom to care for. He transferred most of his men to King Baldwin's command, and many of these Norwegians would spend the rest of their lives in the Holy Land. As a token of his appreciation, Baldwin gifted Sigurd with a splinter of the True Cross – the very cross upon which Christ had been tortured. Sigurd then bid goodbye with his friend and left.

The Return Home

He first arrived in Cyprus, where he stayed for a while before moving on to Constantinople. According to the sagas, Emperor Alexios I was so thrilled over Sigurd's visit that he organised to have the Norwegians enter through the Golden Gates – the glittering arc that the Emperor always used when celebrating his own triumphs. The illustrious Constantinople was decorated in sparkling colours and majestic buildings. The Norwegians were amazed.

Alexios embraced Sigurd and the two spent much time together. Spectacular feasts were arranged for Sigurd and his men, who were well fed and entertained by Greek music, poets and dancers. They spectated the exciting public games in the Hippodrome (chariot racing), letting Sigurd experience the exhilarating spectacles of the Roman world. *"The games themselves are so artfully and cleverly managed, that people appear to be riding in the air; and at them also are used shot-fire* [Greek flamethrowers or firework], *and all kinds of harp-playing, singing and musical instruments,"* the sagas relate.

Sigurd reinforced the Byzantines by giving them all his ships. In return, Alexios provided the Norwegians with thousands of horses so that they could ride easily through the European continent. Many of Sigurd's warriors fell in love with the classical metropolis and chose to stay in Constantinople to serve as Varangians. Sigurd then bid goodbye with the Greeks and left the city.

Many European rulers welcomed Sigurd and his Norwegians on the way home. They met Emperor Lothas of the Holy Roman Empire in Germany and were greeted by Danish King Niels. In 1111, they finally arrived back in Norway. Sigurd's return was celebrated as a triumph. Poets wrote about his deeds and adventures in deep admiration and respect. He became known as Sigurd *Jorsalandsfare*, later abbreviated to *Jorsalfare*, meaning Sigurd the Jerusalem-Farer. In English, he was known as Sigurd *the Crusader*.

Sigurd returned to a Norway that was in far better shape than the one he left in 1108. His brother, King Eystein, had invested all his time and energy in developing the country. He had brokered new trade deals that improved Norway's trading network in Europe. He helped launch the profitable stockfish industry, which later became one of the most profitable trades to ever grace the country and a blessing to many Norwegians in hard times. Salmon was also exported in greater quantities than before, much thanks to Eystein's investments in Lofoten. He spent much time in Bergen, building and administrating, consequently making the city more relevant. He had undertaken major construction projects along the West coast, erecting churches, developing towns and ports. A new network of roads had also been built to boost domestic commerce. Craftsmen and artisans were either imported from Europe or handpicked from society to make detailed wooden carvings in the King's royal hall, town halls, churches, and other monuments. Eystein was therefore a very important King and his contribution to Norway should never be overshadowed by the achievements of his crusader brother.

Sigurd and Eystein talked for hours together, enjoying locally brewed mead, and taking turns to boast about their own achievements. As they began hurting each other's pride, the playful conversation soon turned into a tense quarrel. Eystein explained how he had erected countless churches and made Norway so prosperous that envious Swedes coveted to tax his lands. He affirmed that his actions were "*more to the benefit of the Kingdom*" than Sigurd's crusade. Sigurd, in turn, told a story of how he once swam across the river Jordan and, on the edge of the river, twisted a knot of willows. He teasingly challenged his brother to untie it, claiming there was a curse on him if he failed to do so. Eystein angrily countered: *"I shall not go and untie the knot which you tied for me, but if I had been inclined to tie a knot for you, you would **not** have been King of Norway at your return to this country, when with a single ship you came sailing into my fleet!"* The two brothers went silent after this. Both had hurt each other's pride. Though they maintained peace, they had a contentious relationship.

Troubling Last Years

Olaf, Sigurd's little brother, died in 1117 at the age of 17. It was unfortunate, and it left the power shared between the two elder brothers. Together they implemented the *tithe*, a 10% tax to benefit the Church, and founded the merchant town Stavanger, which would grow to become one of Norway's prevalent cities. However, it was Eystein who remained the effective ruler. Sigurd preferred to retire to his castle where he carefully preserved the splinter of the True Cross and reminisced of his younger days. In 1123, Eystein Magnusson died, leaving power solely with Sigurd.

During these last years, Sigurd struggled with ill health and a troubled mind. *"Many things can change during a man's lifetime,"* he used to say (Morkinskinna). At times, he failed to keep his composure, bursting into rants that startled others around him. Other times, he was sombre and anxious. We must remember that Sigurd had witnessed the traumas of war since he was eight years old. Now, far into adulthood, it all seemed to get an ugly hold on him.

One day, Ottar, one of Sigurd's torch-bearers, stood up in the middle of a holy feast and spoke up against the King. *"Different were the days, Sire, when you came with great state and splendour to Norway, and with great fame and honour; for then all your friends came to meet you with joy,"* he opened, stirring shock and awe in the hall, for it was obviously an outrage for a servant to speak this way to a King. *"...but now days of sorrow are come over us, for on this holy festival many of your friends have come to you, and cannot be cheerful on account of your melancholy and ill heath. It is much to be desired that you would be merry with them, good King..."* Sigurd leaped forward in anger and drew his sword, but as he noticed that Ottar did not defend nor hide himself, he lowered his blade and went back to his seat. The troubled King then publicly confessed his shameful act: *"...I came here like a madman...but he* [Ottar] *turned aside my deed, and was not afraid of death for it. It takes a long time to test the true nature of men. Here sat my most distinguished friends...court officials and the best men in the land. But none was so well disposed toward me as this man, who you* (nobles) *probably regard of little worth."* He praised Ottar for having the guts to speak honestly to his King – the mark of true friendship – and held that none of the aristocrats or nobles loved their King more dearly than Ottar. Sigurd ordered Ottar to be released from servitude and sit with the nobles as the feast continued.

Many in the King's circle probably sought ways to exploit Sigurd's weaknesses to their own gain. His weariness was taking a toll on his decision-making. One day, a man by the name of Harald Gilchrist appeared (*Gilchrist* alludes to *Friend of Christ*. Whether this was a genuine epithet, or a deceptive one, is for the reader to decide). He made a bold claim that he was a bastard son of Magnus Barefoot. Since Gilchrist allegedly came from Ireland, the claim might have been true. But then again, they only had his word for it.

After some debating, Sigurd decided to test Gilchrist's honesty by demanding a trial-by-ordeal. These were very peculiar, medieval trials where the subject had to perform an act of excruciating pain to prove his commitment to the truth. The idea was that no one withstands so much pain over a lie (the only flaw in this case, of course, is that the reward for being considered "truthful" is admission into the royal family, i.e, power, riches, servants, and fame. For Gilchrist's trial therefore, what appears as "commitment to truth" can very easily be pure ambition and self-interest).

Gilchrist had to walk on nine glowing ploughshares to prove his word. He took off his shoes and walked across, but to the amazement of the onlookers, did not faint. This seemed to be sufficient proof for Sigurd, who acknowledged him as an honest man. He embraced him as a brother. In return for being admitted into the royal family, Harald Gilchrist swore an oath never to dispute the power of Sigurd, or the inheritance of Sigurd's son, Magnus Sigurdsson.

In 1130, Sigurd was found dead in his bathtub. It was a mysterious death, and it threw the entire country into a devastating chaos. Magnus Sigurdsson naturally arose as the next King, but Harald Gilchrist immediately violated his oath and disputed Magnus' succession. This was the beginning of the darkest chapter in Norwegian history: the heartless Norwegian civil war.

Legacy of Sigurd and Eystein

Sigurd the Crusader won a legacy unlike most. His Oriental adventures captured the imagination of all Norwegians, as they pictured him valiantly defeating the Pope's enemies and aiding Europe's many monarchs. Sigurd, as mentioned, became the first crusader King in Europe. European rulers warmly embraced him, and he assisted European forces in Spain, Italy, the Holy Land and Greece. He demonstrated for the world that Norway intended to be a great *resource* for Europe – a Kingdom the Pope and the European civilization could rely on. In 1189, the Kings of England, France and the Holy Roman Empire followed in his footsteps as they all embarked on a crusade – the so-called *Kings Crusade.*

It is also important to highlight the constructive life of Eystein Magnusson – the man who made all of this possible by keeping Norway functioning while Sigurd was away. Eystein stimulated growth and prosperity through his excellent management skills. Crucial reforms and projects were accomplished. Eystein gently led Norway to much better conditions, and for that, deserves a cherished name in history.

Sverre

1145 – 1202

"Gentle as a lamb, dauntless as a lion."

– Sverre's motto

Sverre is arguably the greatest military mind Norway has ever produced. His good grasp on strategy, charismatic leadership and quick wit earned him victories upon victories. He usually had the odds against him, but still found a way to prevail. This was a man who perfected the medieval art of war. Sverre also marked a fundamental change in Norwegian history, as he would establish a new rule in the country that would steer Norway towards its greatest extent.

Destruction and divide

"When each one looks only to his own tricks and wiles, great misfortunes of all kinds will come upon the land. Murder and quarrels will multiply; many women will be carried off as captives of war and violated...For then the kingdom was rent, the morals of the people were confused, and their loyalty was divided among a number of lords, each one of whom was striving to contrive and employ against the others cunning, deception, disloyalty, and evil in every form."

– King's Mirror (1260).

The transition of power had failed. The quarrel between Harald Gilchrist and Magnus Sigurdsson eventually caused the latter to muster his troops and march to oust Gilchrist. He defeated him at the Battle of Bohuslan in 1134 and sent Gilchrist into exile to Denmark. However, the Danes reinforced Gilchrist, who soon reappeared in Viken. Having subdued Viken, he attacked Bergen, outmanoeuvred Magnus' troops, and captured the young king. Magnus was blinded, castrated, and mutilated, thus known as Magnus the Blind.

Harald Gilchrist was then sole ruler, but not for long. In 1136, Gilchrist's brother Sigurd Slembe assassinated him and attempted to seize the throne for himself. This threw the country into upheaval: Gilchrist's sons mobilised troops to avenge their father, while Slembe forged an alliance with Magnus the Blind to face them. In 1139, Slembe and Magnus met their opponents at the naval Battle of Holmengra but were decisively defeated. Magnus was slain, and Slembe was captured, facing hours of excruciating torture.

Norway had already collapsed into a state of disarray. The transition of power had completely failed, and it was highly disputed who should be King. Each of Gilchrist's sons now wanted power, and fierce quarrels erupted between them. It did not take long before their feud threw Norway into yet another wave of destructive civil war.

These were horrific years. The custom of blood-vengeance kept striding parties locked in enmity through generations. Shocking atrocities were committed by all parties, and brutality was regarded as the only remedy for enemies. All the while, the Danes kept funding the rivalling parties, probably hoping to keep Norway weak and divided.

The Kingdom splintered into different factions. Wives turned into widows; sons and daughters into orphans. Opportunists aristocrats exploited the chaos to advance their agendas, and angry farmers plotted together to form new uprisings; Norwegians ravaged and plundered each other's towns and cities, simply because they belonged to rivalling factions. The house was divided against itself – how long could it last?

Finally, in the 1160s, the master strategist Erling Skakke assumed power over the country and began to stabilise the realm. Erling was a former crusader with good connections in Rome and Constantinople. In fact, his epithet *Skakke* (meaning *slanted*) comes from a near-fatal neck wound inflicted upon him by an Arab in Sicily, making his head permanently slanting. The key to Erling's solid position in Norway was his special alliance with the Papacy and the Church, and his marriage to Kristin Sigurdsdottir, the daughter of Sigurd the Crusader.

Under Erling's watch, Norway was given some moments of relief and healing. He bestowed various privileges to the clergy, making Norway a *province of the Church* with a diocese in Nidaros. This was important for Norway's expression of sovereignty, as its Church had previously relied on foreign dioceses.

This enabled the new Archbishop of Nidaros, Eystein, to reform Norway's succession laws. He clarified and defined the order of succession to promote political stability. In this context, Erling Skakke placed his son, Magnus Erlingsson, on the throne. This was done to restore the Fairhair dynasty, as Magnus' mother was the daughter of

Sigurd the Crusader. To consolidate the arrangement, Erling and Archbishop Eystein arranged a spectacular, public coronation ceremony. Thus, Magnus Erlingsson became the first King in the North to be *formally* crowned and *anointed* by an Archbishop.

However, this was controversial. Usually, the throne could only be passed down to *sons* of a former King. But with Magnus, it had passed down from a noble (Magnus was tied to the Fairhair dynasty through Princess Kristin, but in medieval tradition, this was insufficient). The audacious arrangement had then been endorsed by the Papacy. Next, they had authored and enforced a new law that only legitimate sons of the (current) King could inherit power. In other words, the son of a noble had become King and then enshrined in law that only sons of a King could become King. Many Norwegians probably felt that this was a slick political coup.

Erling initiated a purge of all remaining suspects and troublemakers – but here lay his cardinal error. As series of local governors were executed, civilians became very upset. Many of these local rulers were loved by the population but were now found drowned or hung, executed on a questionable mandate.

A new rebel faction arose, composed of poor peasants who did not afford proper equipment, but still had the stomach to fight. They were called the Birch-legs (*Birkebeinere*) as they used birch as shinpads. However, at the Second Battle of Re in 1177, they were annihilated and dispersed, with their faction leader killed. Nothing seemed to stop Erling Skakke and his son, Magnus Erlingsson, from complete domination over Norway – but then, a certain Sverre appeared.

Humble beginnings

Sverre was an orphan. He grew up at the Faeroy Islands, far away from these calamities. He was fostered by a bishop and had a theological education, being ordained a priest in his early 20s. When he turned 24 years, his mother Gunnhild came to the Faeroes and, according to *Sverres saga*, informed him that he was the grandchild of Harald Gilchrist and therefore a royal with a legitimate claim to the throne[30].

---

[30] *Sverres saga* was written on the orders of Sverre himself, and the piece is very biased in his favour. Scholars therefore dismiss his connection to the Fairhair dynasty as being invented to make him appear legitimate. Besides, why would a royal child live in the Fareoys and become a priest? Even the Birchlegs doubted his authenticity. On the other hand, maybe we can give him the benefit of the doubt. Technically, we have no evidence that *he wasn't* of Fairhair's kin.

Sverre immediately voyaged to Norway in 1176 to see what he could do. He was optimistic – but rapidly found out that his chances for success were abysmal. After failing to win any supporters, he met Earl Birger of Sweden. Birger disclosed that he backed the unfortunate Birchleg faction, who were utterly disorganised, demoralised, and broken. There were allegedly just 70 warriors left of the Birchlegs.

Together with Birger, Sverre rode into their main campsite, gathered all the men he could and held a flaming speech. As a preacher, Sverre had a gift for lyricism. The Birchlegs must have noticed his fervent charisma and passionate energy – but despite this, they did not find him credible. His story of how he was Gilchrist's grandson was too far-fetched and dubious. Under the influence of Earl Birger however, they reluctantly accepted *Sverre the Priest* as their new leader. After all, they had nothing more to lose.

Sverre was an optimist, but also a realist. He acknowledged the Birchleg's obvious weaknesses, but also noticed some potential strengths. Behind the demoralised faces, Sverre knew the Birchlegs had heart – all they needed was a victory to boost their morale. They were poorly equipped and many lacked sufficient fighting skills, but with a focus on mobility, speed, deception and discipline, Sverre believed he could win. He simply had to avoid open battle, knowing that the Birchlegs were not ready for such a confrontation. With emphasis on guerrilla warfare however, the Birchlegs would gain greater confidence and, one day, be ready to take on the enemy in a full-scale clash of arms.

The Winter Campaign

Eastern Norway suddenly experienced a thunderbolt that shook the comfortable nobility. The Birchlegs took fief after fief, storming through the snow and escaping into the woods at an uncatchable pace. They dashed from fort to fort, overrunning minor defences and looting arsenals and warehouses. They operated with an energy unseen before and it motivated other peasants to join them. Their group expanded. Soon, 100 turned to 500; 500 to 1000, and so on. These rapid attacks gave the Birchlegs much training and experience. Their morale began to surge.

Sverre led his men over the steep mountains of Sogn and attacked Voss, taking the garrison by surprise. Shortly after the attack, the Birchlegs hurried back into the mountains before the royals could retaliate. To stay undefeated, Sverre took a breath-taking gamble by *wintering* with his men on top of the Sogn mountains. As the winter storms swept the mountains, the men underwent great sufferings. *"So utterly worn out were the men that not one was capable of kindling the fire,"* Sverre's saga admits. Amid these terrible conditions, Sverre held sermons for his men, encouraging them to hold on with courage and gratitude to God and quoting Scripture – Sverre was, after all, a priest.

Having survived winter, Sverre stormed to the North and shocked the country by defeating royal troops near Nidaros. When spring returned, he captured the city of Nidaros – the key to Norway. It was to everybody's amazement. The people of Trondelag hailed Sverre as King of Norway and many enlisted to fight alongside him.

An alarmed Erling Skakke realised that it was time to take the Birchlegs seriously. He assembled a vast fleet, disembarked in Trondelag and went on a relentless pursuit for his foe. Sverre resulted to guerrilla tactics once more: he sent squadrons of Birchlegs on skis to engage in short, tactical skirmishes with the enemy, ultimately confusing and demoralising them. Through the many fjords and valleys of the North, the Birchlegs moved like shadows.

One day, Erling caught Sverre's army and ordered a full charge. Sverre signalled a dramatized retreat into the valleys, which convinced Erling that the fragile Birchleg faction had been routed. Confident that victory was his, Erling returned to Nidaros and disbanded his army – but he was fooled.

Sverre turned by Gauldal and suddenly attacked from the North side of Nidaros, by Kalvskinnet. Since Erling had disbanded his forces, the garrison at Nidaros was all he had. The Birchlegs stormed against Erling and cut down his banner, piercing him with a lance. As Erling struggled with his mortal wounds, his son, Magnus Erlingsson, rushed to his side. He saw his father die before him. The rest of the royal garrison panicked and fled. Magnus, too, made a dramatic escape. The Birchlegs had decisively defeated the royals.

The Battle of Kalvskinnet (by Nidaros) in 1179 was an astonishing triumph. Sverre had outmanoeuvred the old crusader strategist, Erling Skakke. People now flocked to the Birchlegs to join them.

The Battle of Ila

Back in Bergen, Magnus Erlingsson gritted his teeth in anger. He was determined to avenge his father and wipe out Sverre and his faction for good. He recruited men from across the country and gathered all the warships he could. In 1180, Magnus and the royal army landed in Trondelag.

This was a serious threat to Sverre. The numerically superior royal army had many advantages over the Birchlegs. Worse still, Sverre could no longer resort to his indirect type of warfare. If he did, he could lose Nidaros and thus weaken his faction. A pitched battle was inevitable, whether he liked it or not.

But Sverre needed more time to prepare for it, so he sued for ceasefire. Magnus was interested in peace, and so began a series of time-consuming peace talks. Meanwhile, Sverre's marshals recruited and trained troops *en masse*. As soon as the army stood ready, and Sverre felt confident he could lead them in a heavy engagement, he aborted the peace talks and the war resumed. It seemed to be a mere deceptive trick to buy time.

An irritated Magnus stormed against the Birchlegs and came to where they had taken up position, near Ila. Sverre had a primitive headquarters here, consisting of a simple wooden fort and a palisade. Seeing the simple design of Sverre's defences, Magnus was confident. His trained, professional soldiers could easily annihilate the peasants Sverre led.

However, Magnus had the terrain against him. He stood on a thin, 200-meter wide strip of land where his numerical advantage was useless. On his flanks, he had a river and a fjord. A frontal assault was his only possible move. Furthermore, the sunlight shone against him and blinded his troops.

Sverre, who had chosen the ground, was in a much better position. He stood on higher ground, had a fort to his back in case of a retreat and a palisade on his left flank. Despite all this, Magnus still believed his superior numbers would bring him victory. He was probably agitated over Sverre's annoying ploys and having finally "caught" his elusive army for a pitched battle, he was eager to attack.

Meanwhile, Sverre sensed that there was a nervous energy among his men. It was their first heavy engagement since their disastrous battle at Re and they were now outnumbered by heavily armed soldiers and barons. Sverre rode in front of his ranks and spoke fondly to them: *"A great host and fine body-guard are here come together, and it now appears that we have to deal with overwhelming odds. The ranks of King Magnus, with gilt weapons and in goodly raiment, cover the whole fields,"* he opened, gaining their full attention as he acknowledged their fears. But he went on: *"...Lo! now, my brave fellows, it is enough to have a choice between two ways-the one, to win victory; the other, to die with honour...in every battle where you are present, either you will fall or you will come forth alive. Be valiant, therefore, since all is determined beforehand. Nought may send a man to his grave if his time is not come; and if he is doomed to die, nought may save him. To die in flight is the worst death of all."* This stoic message encouraged the men. *"Birchlegs! Be sure of this:...there is only one way out: stand firmly fast, and give them no room to advance..."* Sverre then explained his overall strategy to win, re-assuring them that they were in control despite the grim odds. After this, he introduced some humour to lift their spirits: *"Besides, the greatest part of them are more at home at a wedding than a fight,and are more accustomed to mead-drinking than to warfare!"* The soldiers chuckled and cheered. He then gave his last order: *"March well forward now, my brave fellows, and may God have you in His Keeping."* [31]

Soon after, the royal army began advancing towards them. Magnus gave the order to charge, and they stormed at the Birchlegs, with the brunt of their forces concentrating on Sverre's clear and visible banner on the left flank, hoping to kill Sverre. They brawled violently with the Birchlegs, who valiantly held out.

While pinned down on the right, some movement occurred on the right: it was King Sverre and his cavalry! He swung around the royal army, routed its reinforcements, and charged at them from behind. As it turned out, the banner of King Sverre was just a bait, intended to drag the main royal army to the left so they would expose their right.

The shocked royal forces disintegrated. A vast sway of soldiers fled in chaos, but were soon trapped by a delegation of Birchlegs who blocked their escape route. Sverre himself was lethally wounded when an enemy noble hurled his sword at his head, cutting his ear and neck. Fortunately for the Birchlegs, it was not a mortal wound. Sverre's cavalry pressed on and cut the enemy down, one by one. It was a total annihilation of the royal army. Magnus Erlingsson escaped alive.

To the Birchlegs, this was a crushing and historic victory. They were now fully confident in the leadership of Sverre, who had restored honour and hope to their cause. For Magnus, however, it was a humiliating defeat. Funded by King Valdemar II of Denmark, Magnus raised a new warfleet and clashed with Sverre again in the naval Battle of Nordnes, which resulted in a tactical victory, but strategic draw.

At this time, Sverre's Birchlegs were low on resources and needed a moment of respite. Sverre took a risky gamble: he dashed over to Viken and looted the territory to stock up, temporarily exposing Trondelag and Nidaros. Magnus saw the opportunity and hurried North. He captured Nidaros, plundered Trondelag and burnt down *all Sverre's docked warships*. It was a significant blow to Sverre's position.

To restore strength to his faction, Sverre turned to stealth. While initiating a reconstruction of a warfleet, he took a few, smaller boats and sailed into Bergen at night. With the skilful use of agents and diversionary manoeuvres, he hijacked Magnus' *entire royal fleet* from the docks and sailed out. Now with a proper warfleet, he invaded Bergen, occupied the city and drove Magnus into refuge. It was a remarkable feat, involving

---

[31] Quoted from *Sverre's Saga* which was dictated by Sverre himself.

tactics unseen in Nordic warfare. Within just a short period, he went from a crisis to victory.

The Battle of Fimreite – Norway's Defining Moment

King Valdemar II of Denmark welcomed a defeated Magnus Erlingsson yet again. He decided to give the young King another chance. A new and powerful fleet was constructed, operated by vast numbers of soldiers. In 1182, Magnus made a highly concentrated effort in defeating Sverre for good. His army reached Bergen ready to wage total war – but once there, they found no enemy. His agents scouted around but found only the garrison. Where did they all go?

Locals informed him that the Birchlegs were deep into the Sogn fjord, with Sverre residing in Kaupanger. At all haste, Magnus mustered his forces, sailed into the majestic fjord and confronted his nemesis.

Near Fimreite, Magnus' men suddenly spotted a massive warship. This was the *Mariasuda* – a gigantic naval vessel that looked more like a floating fortress than a battleship. It was a maritime innovation of its day, engineered after Sverre's own design. Its height allowed for Birchlegs to shoot arrows down on the enemy from a higher position. Its sheer size also had the effect of pinning down numerous warships, locking them together. Suddenly, some smaller Birchleg vessels appeared from behind, surrounding Magnus' royal fleet. Sverre was on a smaller boat at a distance, observing and directing his troops. He ordered to have his vessels attack in groups of three, so that in each minor engagement, his ships would *always* be numerically superior.

It worked. Casualties spread quickly and the royal fleet struggled to stick together. Ship after ship was cleared – then, Magnus' own ship was wrecked. Magnus Erlingsson fell from the sinking ship and drowned. Sharing his fate was the majority of Magnus' staff, nobles, companions, relatives and friends whom were all on their King's ship. Those who managed to swim ashore were captured.

A few days later, some locals had found Magnus' dead body and brought it to Sverre. He took it to Bergen, where he laid it in a decorated open casket. Many came to see dead Magnus and went away weeping. Sverre arranged a large funeral service and called him, despite all their disagreements, an honourable man, adorned by kingly descent.

With this crushing victory, Sverre had secured the throne of Norway. It was one of the most defining moments in Norwegian history, because it led to a complete recreation of Norway's ruling class. The elite that had thrived since Fairhair's days was gone. A new elite now appeared: the Birchlegs. Sverre "the Priest" had literally taken these men from rags to riches, and at the same time forged a new ruling dynasty.

Reign of Sverre and the Bagler War

Sverre launched a reconciliation policy where he encouraged his loyal subjects to marry women from rivalling factions. He himself married Margareth, daughter of King Eric the Holy of Sweden, who had supported Sverre since the beginning. Margareth borne him three children.

However, the whirlwind of the civil war had not yet settled. Despite his attempts to bring back stability and unity, new rebel groups quickly emerged, often consisting of former supporters of Magnus. Usurpers and pretenders sprung up from various places to challenge the hegemony of the Birchlegs. The feud with Sverre continued.

The first opposition came from Jon Kuflung, allegedly Sverre's cousin. He won rapid support in Viken and in the Westlands but was defeated by Sverre in a surprise attack in 1188. Next, the øyskjeggene ("Isle-beards" or *Eyskeggs*) rose to the challenge. They were a collection of former servants and supporters of Magnus Erlingsson, now claiming the throne through Magnus' infant son, Sigurd. Their army was mostly recruited from the Orkney Islands, hence their nickname. They were formidable fighters and hard to beat. In the huge naval engagement at Florvåg outside Bergen, Sverre and the Birchlegs were almost defeated, but won due to their discipline and battle-hardened experience.

The "red thread" in all these clashes were, of course, the power struggle between Sverre and the old aristocracy. Sverre's birchlegs did not belong to ancestral noble families. They were poor commoners who had suddenly risen to become aristocrats. Though undeniably Christian, they were sceptical to the political power of the Papacy.

The situation worsened when the Archbishop Eystein refused to acknowledge or anoint Sverre as King of Norway - a significant weakness in Sverre's position. The argument soon escalated into a heated quarrel over the fundamental principles of Church and State. In essence, Archbishop Eystein viewed the Church as being both sovereign from and superior to the monarch, on the grounds that any worldly Kingdom ought to be scrutinised by the Servants of the Heavenly, namely the Bishops. In Sverre's view, God and the institutional Church were not the same. The King was a subject of God, *not* of the Pope. The bishops ought to serve the King, not the opposite. Implied here was that Sverre should have the right to promote bishops and priests as he wished, independent from the Papacy. This was a gross trespassing in the eyes of the Church.

Sverre practically sought to reverse the privileges that Erling and Magnus had bestowed on them some decades before. For this, Sverre is often known in Norway as the man who "spoke against Rome".

Other related matters were also drawn up, like the dispute over ownership and management church buildings. Traditionally, the founder of a Church became its owner and manager – but with recent reforms, the Church seized ownership and management, regardless of who the founder was. This was very controversial and upset a lot of people, especially as it sacked them of valuable income and labour. Sverre voiced this sentiment quite clearly: *"They demand that we build churches, but once built, we are chased away like heathens!"* He continued, sarcastically: *"we are certainly allowed to deal with all the financial costs, but we mustn't rule over what the costs were for!"* His quarrel with the clergy became so loud that the Archbishop himself fled the country.

In 1190, a compromise was reached and Sverre was finally crowned King of Norway – but the whole affair had already caused irreversible damage: the Pope excommunicated him and raised interdict over all peoples who supported him. Following this, a hostile Church rebellion was unleashed. Bishop Nikolas Arnesson of Oslo led the so-called *"Baglers"* to arms. They swiftly had control over Viken and opened a new front against Sverre.

These men were fierce and well supplied. After tactical victories, but strategically indecisive results for Sverre, he retreated to Bergen – exposing Trondelag to the enemy. In 1197, the Baglers invaded and captured Trondelag, dismantling Sverre's fortresses there and forcing his subjects to surrender. In the following year, Sverre attempted to retaliate, but failed to surprise his enemy and had his army caught in a disadvantageous position, leading to a sound defeat. He limped back to Bergen as the momentum was now with the Baglers. They pursued him and laid siege to Bergen – the capital. After a seemingly endless period of ghastly fighting around the fortresses, the Baglers deliberately set fire to Bergen, burning down houses, supply depots, and, reportedly, churches. With most of his stocks destroyed, the Birchlegs starved and Sverre was forced to abandon the capital. He went North. By this time, the Baglers maintained control over the West, South and East of Norway. All he had worked for was slipping into the hands of the enemy. In this dark, desperate moment, Sverre received news that his eldest son had died of disease. It was a disheartening chapter in his life.

But Sverre had enormous willpower and refused to accept defeat. He arrived in Trondelag and, due to his captivating speeches, gained vast numbers of volunteers. The Tronds had a special, unfailing devotion to Sverre. He was a Nordic wolf, guilty of all kinds of mischief and transgressions – but he was genuine. Few believed his legitimacy as King – but, in a way, it didn't matter. He appeared as a rugged, brutally honest man, who represented self-rule and independence. Sverre also received backing from England and Sweden, who reinforced him with supplies and elite mercenaries, like English Longbowmen.

With a smaller fleet and a weaker army, Sverre met the Baglers at Strindfjorden in Trondelag. It was a long, unforgiving battle – but ended in a much-needed victory for Sverre. Subsequently, he took full advantage of the new momentum: chasing the fleeing troops out of the Trondelag, reclaiming lost land and invading Viken. Viken was, however, the heartland for Bagler support. During the winter 1199-1200, thousands of enemies surrounded him, making defeat seem inevitable. But Sverre and his experienced fighters withstood the calamity, and, division by division, drove their numerically superior foe into disintegration. Oslo was taken. Next, a gruelling and exhausting siege took place at Tonsberg fortress, where the Baglers surrendered only after four, harsh months of starvation and disease. With the fall of Tonsberg, the Bagler faction began to look frail. Victory, yet again – or so it seemed.

After all of this had been accomplished, an exhausted King Sverre fell ill. He had persevered tens of extreme battles, survived a lifetime of protracted war and enjoyed only sporadic moments of peace. It all took its toll.

The 51-year-old King rested in Bergen while being accompanied by his dear comrade Reidar. According to the sagas, Sverre spoke with Reidar for hours, as Reidar was known to be an exceptionally wise man. Sverre stressed the need for reconciliation with the Church and hoped that his second son and heir, Haakon Sverresson, would do so. Only this could produce lasting peace. But the illness proved too strong for him. *"The kingdom has brought me labour and unrest and trouble, rather than peace and a quiet life,"* he said to his marshals on his deathbed, *"...let my Lord now judge between me and them, and decide all my cause."* He died on March 9th, 1202.

Legacy

"He was most eloquent in speech; his ideas were lofty, his articulation was distinct, and when he spoke, the ring of his voice was so clear that though he did not appear to speak loud, all understood him...He never drank strong drink to the injury of his reason, and always ate but one meal a day. He was valiant and bold, very capable of enduring fatigue and loss of sleep...[he was] very alert at night."

– Anonymous but quoted from the post-mortem
pages of Sverre's saga.

Who would have known that an ordained Priest from the Faeroys would become such a warlord in Norway? Sverre the Priest became one of, if not the, greatest military commander in Norwegian history. He demonstrated exceptional knowledge of the arts of war – perhaps he was a student of Caesar's. He transformed a broken, demoralised, and poor group of men into an elite force and led them to stunning victories again and again.

Sverre's warfare carefully used advantages in nature and terrain, psychology and deception, spies, and agents, to prepare better winning conditions before engagement. His disciplined soldiers were drilled in tactics and strategies, were bound by comradery, and trusted their commander. They moved with speed, attacked with decisive force, and manoeuvred with elusive mobility – making them highly effective units for both pitched battles and long campaigns. Sverre mastered the balance of knowing when to fight, and when not to fight. He was a realist, acknowledging the limitations of his men and supply lines; but he was also an optimist, knowing how far he could push them to make them realize their full potential. He never gave them a command he knew they could not execute; and never let them slack when they ought to go further.

The warrior-priest also introduced new ways of waging war in Scandinavia. He broke with the Norse tradition of fighting on the front-line, preferring instead to command his troops from a high ground where he could view the field. He also demonstrated the effectiveness of heavy cavalry at the Battle of Ila; perfected the use of skis for military mobility and used them to harass enemy lines; and introduced more sophisticated artillery, using battering rams and Roman-style ballistae that had never before been deployed in the North. His understanding of engineering helped bring new ideas within fortress architecture, and the multiple *"Sverre-forts"* across Norway featured a special design that would become very popular. Many of these forts still stand today, notably in Bergen and Trondheim (Nidaros).

Personally, Sverre was full of energy and extremely hard-working. He was an eloquent orator and gripping public speaker, electrifying the men before battle or during arduous times. He is described by his contemporaries as calm, even-tempered, true to his word, magnanimous but also fiercely ambitious. He was very cautious in choosing friends and could endure a lot of hardship. He was courteous to his friends, but hard on his enemies. Sverre was adored by his Guardsmen and soldiers, as he was always *"hopeful under misfortune."*

However, Sverre was also a strict disciplinary. The men were scolded if they were caught sleeping on guard or complaining. In his reprimands, Sverre often referred to the *old Birchleg veterans* – the toughest bunch of fighters he had ever met – who would never complain. If they were caught mistreating civilians, they were severely punished. Overdrinking was also not permitted. Sverre saw overdrinking as a plague in society that needed to be curtailed. He lectured his men about it, explaining how it wasted their wealth, destroyed their health and mind and roused them to reckless or unrighteous deeds. *"When all wealth, health and reason are destroyed,"* he concluded, *"it incites a man to destroy what is not yet lost, his soul."* All men ought to treat townsfolk, merchants and women with respect, he bid them. *"Warriors in time of peace should be gentle as lambs, but in war dauntless as lions."* His men were captivated by his sermons and listened carefully. *Sverre's Saga* boasts towards the end: *"It is of common opinion that there has never been better warriors in all of Norway than the men who followed King Sverre."*

Sverre fundamentally transformed Norway. He had elevated the poor Birchlegs to become the new aristocracy. Many – both medieval and modern - dismiss his spurious claim to the royal bloodline as invented, making him illegitimate and therefore not a part of the Fairhair dynasty.

Thus, his reign is often marked as the *end* of the Fairhair dynasty and the *beginning* of the Sverre dynasty. This new rule would serve as the platform for Norway's second golden age. Sverre's descendants had a great future ahead of them.

Haakon Haakonsson

1204 – 1263

As the Kingdom was at the brink of collapse, Norway desperately needed a King who could end the destructive civil war and restore peace and prosperity. That man was Haakon Haakonsson. He had the character and integrity all Norwegians could rally behind to end the civil strife, and he had the vision and grandeur to lead them to their zenith. Not only did Haakon Haakonsson rebuild Norway – he made it mightier, larger, and more prosperous than it had ever been.

A Hunted Child

Haakon Sverresson, King Sverre's son, took over the Kingship of the Birchlegs. He answered to his father's last wish and sought peace with the Church. The bishops who had been exiled from Norway under King Sverre were invited back for negotiations. Haakon Sverresson gave in to most of their demands, restoring Church power and influence. For this, Norway was relieved of its Papal excommunication and interdict and the Church ceased their relationship with the Baglers.

It is no wonder, then, why the people praised Haakon as a man of peace. He took time to treat the commoners fairly and with dignity. The population began rallying behind the Birchlegs and the Baglers gradually lost support. Unfortunately, however, Haakon Sverresson only reigned for one year, as he tragically fell ill and died in 1203, possibly of poisoning. He had no obvious heir, leaving the Birchlegs leaderless and reviving the Baglers for war.

However, Haakon actually had a child. Inga, a mistress of Haakon, gave birth to his son in Viken in 1204, who was also given the name Haakon. Since this child was in theory the most legitimate heir to the throne of the Birchlegs, he was a serious threat to the Baglers. The Baglers went on a rampage in towns and villages, searching for the infant. In the hour of desperation, the mother Inga gave the child to two skiers called Torstein and Skjervald, pleading them to bring him safely to Nidaros. The skiers accepted the mission and set out into the wilderness, hastily pursued by hunting Baglers.

An unexpected snowstorm swept across the mountains, but the skiers still managed to push their way through the rough terrain and reach safety.

By the time Haakon entered Nidaros, Inge Baardsson, a relative of Gilchrist and son of Sverre's (supposed) sister Cecilia, had risen as the new King of the faction. Renewed hostilities with the Baglers had almost driven Inge and the Birchlegs into defeat but culminated instead in a peace treaty in 1208 where the country was divided in two. The Westlands and Trondelag to the Birchlegs, and the Eastlands to the Baglers. It was a fragile ceasefire.

Haakon Haakonsson grew up under the watchful eye of his mother Inga, his foster-father Inge (a huscarl) and spent a lot of his childhood with Earl Haakon the Mad, a lord in the Birchleg faction. The Birchleg soldiers loved the young lad and would often play with him. They were Sverre's veterans, and through their love for Sverre, they loved little Haakon.

Haakon received a full academic education by the Cathedral School in Bergen, and later in Nidaros. He is described as remarkably clever. He enjoyed singing but was told to stop as psalms were fit for priests, not for future Kings. Haakon was humble and often quiet, but very resolute. He often amused the Birchlegs by how mature he was for his age. His conversation with the huscarl Helge perfectly exemplifies his cleverness. When Helge disclosed the news that King Inge and Earl Haakon the Mad had made an agreement to exclude Haakon from power, Haakon confidently stated: *"...I doubt their arrangement will hold, as there was no one there to answer on my behalf."* Helge asked who this attorney would be, to which Haakon replied *"God and Saint Olaf, in their hands I rest my case, and they will surely see to it that I get my share."* A proud Helge embraced the young prince. *"Well spoken, young Prince. Such things are better said, than unsaid."*

The Last Rebellions

In 1217, King Inge of the Birchlegs died, causing much debate over who would inherit leadership of the faction. At the following Thing, veterans from Sverre's time flocked around the 13-year-old Haakon, hailing him as King - but there was opposition.

The prominent Earl Skule, half-brother of Inge, held enormous influence among the nobles and instead supported Inge's son, Guthrum. Haakon's supporters voiced their opinions fiercely: *"...we will not serve any other King, than him* [Haakon], *and we will risk our lives and all our possessions for him and that he wins his rightful inheritance... we will oppose every other person you claim as King."* Their loyalty to Sverre's family was unyielding. It was through Sverre's family that they had gained status as aristocrats. If that lineage died out, their newfound status would also be at risk of being eroded.

After a hefty debate, Earl Skule wisely decided to back down and acknowledge Haakon as the legitimate successor of the Birchleg throne – but not without compromise. Skule also secured himself a set of powerful privileges. He was granted rulership over Trondelag, awarding him enough influence to effectively rule as a *de facto* ruler as long as Haakon remained a child.

Later that year, King Philip of the Baglers died childless, leaving the Bagler faction leaderless. Both Skule and little Haakon rushed to Viken to take advantage of the situation. Here, Skule demonstrated his excellent negotiation skills. He convinced them that the Bagler faction was simply obsolete. The Birchlegs had reconciled with the Church, so there was no reason for the Baglers to keep resisting. Besides, since they had no strong leader, they had no other choice than to reconcile with their enemies and accept young Haakon as King. It made sense. By 1218, the Bagler faction officially dissolved.

However, many Bagler soldiers refused to stand down. The feud against Sverre's house was old, but not dead. The grievances ran deeper than the politics between the Birchlegs and the Papacy. The grievances were personal, and unforgiving. Many still yearned to "settle the score" with the Birchlegs.

In the vacuum left by the Bagler dissolvement, new uprising factions sprung up. The most dangerous of these were led by Sigurd Ribbung, allegedly the grandson of Magnus Erlingsson, Sverre's historical nemesis. Scores of old Baglers flocked around Ribbung, who then led them to regain control over Viken. These became re-named the *Ribbungs*, after their leader. It was clear that the curse of the civil war had not yet lifted.

Sigurd Ribbung had an intimidating talent for war. With fast and flexible manoeuvres, quick and decisive shock-attacks and well organised and experienced Bagler units, he routed *everything* Haakon sent towards him. The Birchlegs were simply not as efficient as before. What was once a highly fast and deadly military, had become a lumbering beast. Their armies were now large and slow. Their leaders were rich. Meanwhile, the Ribbungs were the underdogs. It was now *their* turn to be the fast and deadly. It became so disastrous for the Birchlegs that Earl Skule had to personally intervene.

Haakon reached maturity in 1223, and so the status of his Kingship had to be discussed again. Many already doubted his authenticity, or his ability to rule independently from Skule. Moreover, with the rise of new pretenders like Sigurd Ribbung, there were other alternatives.

A vast Thing gathering was held at the Gulathing: landowners, soldiers, noblemen, clergymen – all men of significance showed up to discuss and decide on the fate of Norway. Among the candidates were Haakon, Sigurd Ribbung, but also Earl Skule, who put forth a personal claim to the throne. He argued that he was the former-King Inge's half-brother, therefore nearest kin and rightful heir to his inheritance. A young, but capable, Haakon rose up to respond: *"Sure, you are heir to King Inge and all his inheritance – but this is not the Norwegian throne! King Inge was appointed only to guard my inheritance for me!"* His message was clear and true: Inge was never a part of the Sverre dynasty. He had been an outsider who was placed as a temporary monarch until Haakon reached maturity. Haakon was now mature, so Inge's inheritance was redundant.

Veteran Birchlegs from Sverre's time advocated fiercely for Haakon: *"Us old Birchlegs endured the greatest travail alongside King Sverre, and most of us spilt much blood,"* their spokesperson opened. *"And we thought we wouldn't have to speak of this any longer, because all of us fought to save his inheritance and that of his descendants!"* People rose up and shouted accusations back and forth to demean their rivals. It was a heated scene – but in the end, the bishops and priests, and the majority of the state servants, nobles, veterans, military commanders, legal experts, landowners and farmers found consensus by proclaiming Haakon Haakonsson the fully legitimate King of Norway.

War with the Ribbungs resumed and the Birchlegs lost major territories in the Eastern regions. Earl Skule, probably embittered by the outcome of the Gulathing, refused to aid Haakon. The course of the war was so bleak that many aristocrats now tried to persuade Haakon to divide his realm between him and Sigurd Ribbung. Even the Church argued along those lines – but Haakon adamantly refused.

A few years later, his patience and stubbornness paid off. Sigurd Ribbung died unexpectedly, and the new leadership completely lacked his talent. At the Battle of Vorma the Birchlegs won a decisive victory. By 1227, the Ribbungs surrendered by King Haakon. To promote reconciliation, the King forgave them.

A jealous Earl Skule was then the last uneasy element in Haakon's rule. Their relationship worsened by the day. Ill rumours spread between them, causing misunderstandings and frustration. In 1235, Skule trespassed into Haakon's territories, nearly provoking war. Trying to preserve peace, Haakon married Skule's daughter and gave Skule the prestigious title of Duke, but apart from these formalities, nothing changed. Four years later, war broke out yet again as Skule rebelled. Many former Ribbungs rallied behind him.

Haakon lost the first battle, severely discouraging his men. Upon hearing their complaints, Haakon held a resounding speech: *"in the old days, this wasn't much to whine about!"* he scolded. The King refused to quit. He boldly sought battle anew, this time attacking Skule's positions in Oslo. Heavy fighting occurred here and Haakon took personal command of several divisions. His numerical superiority overwhelmed Skule, who fled the field.

Skule fled to Nidaros to recruit troops, but his defeat in Oslo ruined his reputation and confidence. He fled to a monastery. Scores of Birchlegs were in hasty pursuit. Once they located their nemesis, they torched the monastery, choking Skule to exit. *"Do not strike me in the face, it is not dignified to do so with lords,"* a sentimental Skule pleaded. These were his last words.

Haakon Haakonsson was then the undisputed, uncontested King of Norway. After so much spilt blood, weeping, destruction, terror and pain, the civil war had finally ended. The year was 1240.

Securing the Throne

For the first time in 110 years, all of Norway was under undisputed authority of one King. It was an unprecedented opportunity to end the destruction that had ravaged the country for more than a century. Norway had been in constant insecurity: towns and villages had been plagued by raids from various factions; local governors and nobles had been removed or executed for being on the wrong side of a conflict; many families had lost their spouses and relatives; and many children born into a civil-war mentality. It was a war-torn country that Haakon ruled in 1240.

The first and urgent task for Haakon was therefore to fully secure his own rule. Many European monarchs had already recognised him, but nothing was promised until the Pope did. He had to fully reconcile with the Church. He sent Papal commissions to Rome, but Pope Gregorius IX ignored his requests. To please the Papacy, Haakon granted the Church greater autonomy in domestic issues and rural affairs in Norway. He vowed to fight a crusade against the pagan peoples of the Baltic, perhaps even confront the invading Mongols who had devastated much of Eastern Europe. The Karelian people, who fled the Mongols, had their asylum accepted by King Haakon, who then converted them. Such news pleased the Papacy. In 1246, Pope Innocent IV officially recognized Haakon. He sent Cardinal William to Bergen where a ceremonial coronation took place.

Haakon became one of the very few Norwegian monarchs to be anointed and crowned by a Papal Cardinal; we may quote the words of Cardinal William himself: *"Now your King is crowned, and he is given such a great honour that has not been given to any man before in Norway."* What the Cardinal then said is even more interesting: *"I was told that I would not see many humans here, and those I would see would behave more like animals than humans. But I see countless humans here, an army of people from this land, and I find their behaviour good. I also see scores of foreigners and so many ships...I have never seen so many ships in a port before, and I would think that almost everyone has arrived with a ship stocked full of goods."* People in Rome had told William that he would barely eat, or that the food of the Northmen was inedible. He was prepared to thirst for days, but instead he found *"all good that exists."* This also implies that despite the turmoil of the Civil War period, Norwegians had persevered and found ways to continue developing their country. They had built numerous churches and impressive buildings, and maintained commercial trade lines with their neighbours. But the instability and wars had also alienated Norway from other parts of Europe, who would view these violent Northeners as crude savages. It was time for Haakon Haakonsson to correct their impression.

Restoring the Realm

Haakon's grand undertaking was to establish an undisputable legal framework that removed all ambiguities on the Norwegian state structure and its laws of succession. This was crucial. A major cause of the civil war was exactly the absence of clarity regarding the hierarchical system and succession laws. This uncertainty had plagued Norway since the Viking Age. So, Haakon consulted lawyers and scholars, drafted new formulations, and put it all in writing.

In 1260, King Haakon publicised *"Haakon's New Law"* – a written legal code that reformed the Norwegian social hierarchy and justice system. It included the 1163 law on succession (from Archbishop Eystein and Erling Skakke) that only sons born within wedlock could inherit the throne. Should the King be childless, his nearest male relative would succeed as monarch. As a result, succession procedures were automated, and no one could dispute them. It was a ground-breaking achievement. Out of all the Northern European kingdoms, only Norway enjoyed an automated system of succession. Norway would never see a royal pretender again.

However, there was more to this lawbook than the topic of succession. It had another central element to it, namely the idea and practice of public justice. An offense was no longer confined to the private sphere – it became an offense against the State. Adding to this, an offender was now held personally accountable for his crimes – eradicating the Norse tradition that he or she could send friends or family to speak in his case. Haakon also made efforts to ban blood feuds and mutilation. Anyone who slept with another man's wife could face exile or be outlawed. Rapists were to be instantly outlawed. Even *attempts* to rape were punished by heavy financial repercussions.

It was also crucially important that young crown-princes were well educated in all matters pertaining to Kingship. The last thing Norway needed was a tyrant. Thus, on Haakon's orders, the *King's Mirror* (Kongespeilet) was written – an educational script intended to be studied by aspiring Kings, but was encouraged to be read by common folk too.

This medieval manuscript takes form as a discussion between a father and son. Their conversation covers astronomy and science; business and trade; theology and moral questions; how to behave in various contexts; and more practical advice on diplomacy, the royal court, courtesy, and chivalry. It's advice on morality were deeply rooted in Biblical proverbs. If young princes would study this book carefully, then hopefully they would be wise and conscientious in their leadership.

King Haakon spent time consolidating and expanding the authority of the monarch. He established a royal secretariat, led by a Chancellor, and implemented several new titles. Symbolic clarifications were also needed. In *King's Mirror,* the King is presented as a servant of God, a steward of God's lands. God has given the King worldly power, so it is the duty of the King to serve God with holy Wisdom – for he will one day answer to his actions to God Himself on Judgement Day. His subjects, however, ought to follow their King with loyalty and not turn to pride, jealousy, or greed, as these are the seeds to moral decay, and moral decay is the seed to the total destruction of society. This was the core of Haakon's philosophy. From his *King's Mirror* manuscript, Haakon personally appears as a deeply religious man.

Sturla Thordarson, an Icelandic chief, legal adviser, and author, wrote: *"He* (Haakon) *focused more on strengthening God's Christianity in Norway than any other King since the days of St. Olaf."* Sturla referred to the countless churches, chapels and monasteries, spanning from the far North of Troms to the most South of Agder, which were constructed on Haakon's orders. He administered the erection of numerous hospitals across the country, supplying them freely with land and food. *"...he was friendly to the poor and needy,"* Sturla added. Indeed, such qualities helped raise the Norwegian society out of depravity.

Haakon had a passion for European culture. As the first Norwegian monarch with a formal education, he wished to bring Norway closer to European civilisation. He organised the construction of monumental buildings in the major cities of Norway, based carefully on traditional European design. He rebuilt and bolstered old fortresses, while also founding new ones. Perhaps the most famous of all his constructions was the *Håkonshallen* in Bergen – a large royal estate where the aristocracy, clergy and the King would discuss state affairs, host official ceremonies or feast together.

His interest in European culture could be traced back to his childhood passion for literature. He grew up with stories of legends like King Arthur and Charlemagne and ordered to have European classics translated to Norwegian. This made stories like *Tristan and Isolde*, or the memoirs of Julius Caesar, more widely accessible to the aristocracy, and exerted enormous cultural influence.

Haakon also had religious texts translated from Latin to Norwegian, like his wife's favourite *Visio Tnugdali*. Such translations prompted contemporary authors and poets to begin writing sagas to preserve their own history and create their own classics.

It was at this time that Haakon's huscarl, Snorri Sturlason, wrote *Heimskringla* – the sagas of the Norwegian Kings. Other famous sagas were also written during Haakon's reign, like *Fagrskinna,* most of the *Islendinga Sagar* (a collection of 43 sagas), and the *Snorra Edda* (Prose Edda). This triggered a wave of saga-authoring that continued after Haakon's age, leading to the creation of more famous sagas, like *Flateyjrabok.* The author of this book owe thanks to Haakon Haakonsson. Had it not been for Haakon's passion for literature, this book would never have been written, as the old stories would have never been captured through the sagas.

Northern Hegemony

Apart from being a proficient legislator, Church-builder and cultural promoter, Haakon was a shrewd geopolitician. In the arena of Northern Europe, he wrenched Norway from foreign involvement and led the Kingdom to become mighty and influential. This was the awakening of Norway's golden age.

During the 1240s, Denmark collapsed into the chaos of civil war as the sons of Valdemar II wrestled for power. Denmark weakened. Both Norway and Sweden saw opportunities here. How could they take the best advantage of the situation? One of the major consequences of the civil war was the anarchy erupting in Kattegat and the Baltic that disturbed the valuable trade routes from Germany to Norway. Many Danes began looting Norwegian ships headed to Rostock and Lubeck, incurring considerable financial damage. The Norwegian merchants were outraged and complained to Haakon, demanding revenge. Haakon however, wisely awaited the situation. If he were to retaliate, the timing had to be right.

✳ Haakon Haakonsson ✳

He contacted Emperor Frederick II of the Holy Roman Empire, Denmark's southern neighbour and the overlord of the trading cities of Rostock and Lubeck. Haakon sent him an abundance of gifts and many letters, tying a close friendship to the German Emperor. This was risky – for Frederick was at the time an enemy of the Pope. However, a charismatic Haakon somehow managed to satisfy both sides of the rivalry. While being Frederick's close friend, he still charmed the Pope and remained a favourite of the Papacy.

One of the most important goods for import was grain, and due to the disorder in Kattegat, grain imports took a serious slump. Haakon looked West to compensate. He frequently wrote letters to the King of England, Henry III, also becoming his dear friend. The King of England was one of the richest and most influential in Europe, making Haakon's alliance with Henry a victory on its own. They successfully organised a trade route between England and Norway, finally supplying Norway of much-needed grain. Haakon later organised a trade route between Lubeck and Norway, boosting Norway's grain supply even further and opening up Bergen for German investments.

Haakon's leadership was already becoming quite noticeable in Europe. News spread of Haakon's military successes, his recent alliances, and the end of the civil war. Bearing a reputation for being a skilled naval commander, European monarchs requested his military assistance. For instance, King Louis IX of France planned the 6th crusade and invited Haakon to join him, offering him the command of the entire crusader fleet. Further South, Alphonso of Castille planned an invasion of Morocco and urged Haakon to join, again offering the command of the fleet. Alphonso's invite was made even more attractive when the Pope himself assured Haakon that an expedition to Morocco would be equally regarded as an expedition to Jerusalem. To forge an alliance, Haakon married his daughter, Kristin, to one of Alphonso's brothers. He entertained the idea of joining Alphonso, but for now, he had other issues to take care of.

The Swedes called Haakon to answer for the violence he had caused in Varmland, some years back. Sweden's King, Eric the Lisp 'n' Lame, was not bluffing; he demanded an apology. Haakon however, replied by affirming that the Swedes had also caused *him* much harm – what they needed was not an apology, but a treaty between Norway and Sweden. In other words, he convinced the Swedes that they were on levelled ground. Both sides arranged to meet at Göta river to discuss the treaty.

For the first meeting, Haakon did not show up. For the second meeting, he showed up – but in the most majestic and intimidating fashion possible. Haakon had constructed a massive royal fleet consisting of 350 warships, all decorated with weapons and frightening symbols. Champions of the fleet were the gigantic *Olafssuda* and *Mariasuda*, the latter inherited by his grandfather Sverre. Ahead of this splendid fleet, Haakon came sliding into Göta river. King Eric lost his nerve, fearing that Haakon came to wage war. He fled the meeting and sent Earl Birger to meet the Norwegians instead. Earl Birger, an experienced and competent man, waited impatiently for Haakon to arrive, and once he did, complained over Haakon's flamboyant display of force. Haakon assured him that he had nothing to worry about – at least not for now. After a few bids and offers, the two reached consensus: a peace treaty and a military alliance against their common foe, Denmark. To seal the deal, Haakon ensured to have his son, Haakon the Younger, marry Earl Birger's daughter, Rikissa.

Now the timing was right. The year was 1256 and Christopher, son of Valdemar II, recently won the throne of Denmark. In the wake of his rise, Haakon laid claims on the Danish region of Halland in compensation for all financial losses suffered by Norwegians from Danish piracy. In coordination with Earl Birger, Haakon attacked and rapidly occupied Halland, holding it until Christopher paid compensation. Disappointingly, Earl Birger failed to show up and provide his promised support, but it did not matter – the revenge of Norwegian merchants was finally enveloping.

Christopher refused to pay compensation, so Haakon sent in his divisions to harass Danish lands. When Christopher *still* refused, Haakon took the matter in his own hands. He sailed into Copenhagen ahead of 310 warships and docked in the capital for days, aiming to demoralise the Danish people. It worked. Christopher finally complied. He promised to pay all sums of compensation and *"forgave"* the Norwegians, meaning he swore never to seek revenge. He signed a humiliating treaty, seemingly on Haakon's premises, as it made Haakon *"the godfather"* of Christopher.

With all this accomplished, Haakon changed policy. He now appeared as a generous, amiable friend, inviting Christopher to feast with him and flattering him by offering the astonishing ship, *Olafssuda*, as gift. Next, he strategically proposed a marriage between his son, Magnus, and Princess Ingeborg of Denmark - a token of his good will and a symbol of the new and everlasting friendship between Denmark and Norway. Magnus and Ingeborg married in Bergen.

We see here Haakon Haakonsson's geopolitical cleverness. He used an old grievance to demonstrate his raw force and prove his superiority. Yet, immediately after, he ensured to restore Christopher's confidence and tie an alliance to seal the new settlement.

✷ Haakon Haakonsson ✷

Christopher was deposed of soon after, but it did not matter – Haakon's son was still strategically married to Princess Ingeborg. As a result, by 1257, Haakon had two sons who could potentially inherit the crown of both Sweden *and* Denmark. The following question is inevitable: was he planning to create a Norwegian-led Scandinavian realm? The roots of a Scandinavian union were certainly sowed here.

The Kingdom of Norway began to act as the superior power in the North. Haakon had an extensive network of international relations that no other northerner ever had. For instance, his letters even reached the eyes of the Sultan of Tunis, who, in addition to the ink on the paper, also received hunting falcons as gifts (taken from the famous Smøla and Edøy islands in Norway) [32]. One could also find Norwegian couriers at the court in Cairo, delivering numerous messages to the Fatmids.

Around 1250, Haakon complained to Emperor Frederick II over recent thefts of Norwegian goods in Lubeck. The Holy Roman Emperor, as Haakon's friend, therefore freely *offered* the entire city of Lubeck to Haakon – but right before they could conclude the deal, Frederick died.

Frederick's passing left the Imperial Crown vacant. This was the most prestigious crown in Europe. It had once prided Charlemagne's head, and at Haakon's time, governed vast territories from central France to Estonia, Sleswig to Naples. The crown was also interlinked with the Papacy, as every Holy Roman Emperor theoretically served as the protector of Rome. It was therefore ironic that the Pope had tried to depose of Frederick, whom he distrusted. With Frederick dead, the Pope reportedly offered the Imperial Crown to Haakon. Europe held its breath. Haakon, however, declined. It was baffling, and scholars to this day wonder why. It may have been Haakon's mistake – but then again, we will never know the intricate details of the situation.

Haakon would rather go on to carve another Holy Roman Empire in the North. He had a grand vision – uniting all lands with Norwegian heritage under one realm, a North Sea Empire. This would fulfil the ancient dreams of Harald Hardrada and Magnus Barefoot.

---

[32] These falcons were most likely taken from the massive falcon industry on Smola (Smøla) and Ed island (Edøy) in Norway. Falcons were an extremely valuable commodity in medieval Europe, and Norway was a top exporter. Though some were sold commercially, most falcons were sent as personal gifts for diplomatic reasons. Smøla and Edøy still have a vast birdlife, and the world's most dense population of sea eagles.

The Norwegian Empire

His first move was to expand Northwards towards the wealthy Sami people – ancient natives of the Arctic, rich on animal furs, hides and ivory. For centuries, Norwegians had traded with them, but their attempts to tax the Samis were always halted by the Princes of Russia. However, by 1250, the Russians were embroiled with chaotic fighting against the Mongol hordes, so Haakon seized the moment to pressured them to give up their claims on the Samis. Alexander Nevsky, Prince of Novgorod, dropped the claims. Thus, Norway's taxable territories now stretched as far North as Finnmark – giving Norway lands it holds to this day.

To the West, Haakon consolidated Norwegian authority on the British Isles, on Shetland and the Faeroes. Iceland was a tougher knot. The island was still ruled by rivalling chieftains. In the early 1220s, Haakon extended political influence to the Icelanders, trying to persuade them to accept a King, but persuasion failed. Ever since, Haakon had taken a different approach. He began hiring Icelandic chiefs as his personal huscarls. This gave them a competitive edge in Icelandic politics, but also made them servants of the King. If a chief quarrelled with Haakon, he could be lawfully prosecuted as mutineers and easily replaced by another chief – for example, his rival in Iceland. Haakon played on these inner rivalries to slowly expand his hold on the chiefs.

One by one, the chiefs surrendered their lands to Haakon, and in turn, Haakon made them governors and bestowed upon them prestigious honours. By 1263, this process was complete. Haakon went ahead to promote *one* man as Earl of Iceland. All of Iceland now taxed to the Norwegian crown. By then, the Greenlanders also formally submitted to Norway.

What remained was Norway at its *greatest extent*. The Norwegian realm spanned from the snowy plains of Finnmark to Agder, from Göta river to the Hebrides and Isle of Man, from the Orkneys, to the Shetlands, Fareoys, Iceland and even as far as Greenland. This widely stretched realm created a lucrative network of trade routes, connecting all of the North together and sending galleys of precious goods back and forth. Bergen became the trading hotspot of the North Sea, working as a magnet that pulled traders from across Northern Europe closer to Norway. It was the *new North Sea Empire* – the Norwegian Empire.

Clash with the Scots

This territorial expansion worried Norway's neighbours. Norway's consolidation of her British islands clashed with the interests of Scotland. The days of Magnus Barefoot were long gone, and Scotland wished to reclaim their lost territory. King Alexander II of Scotland therefore laid claims on the Hebrides, repeatedly offering to purchase them from Norway. Upon the first bid, Haakon taunted the Scots with sarcasm: *"I was not aware that I am so poor that I need to sell land for gold."* He consistently refused all offers. Alexander III, the next Scottish King, took a much more aggressive path. He attacked the Hebrides, triggering the Scottish-Norwegian war.

King Haakon mustered 120 warships and sailed to Scotland. For a moment, Haakon's armada was halted as the Scots sued for peace – but after losing valuable time, the Norwegians realised it was all a deceptive trick to delay the offensive. Haakon resumed the campaign in September. He sailed around Scotland and approached North Ayrshire – but then, heavy Autumn storms hovered over the waves and harassed the Norwegian fleet. Some ships were dispersed, and the army demoralised. Once the storm settled, they went ashore near Largrs to recover and rally – but while there, the main Scottish army appeared.

A smaller Norwegian force was on top of a hill, while the main force was by the beach. The smaller force ran down the hill to merge with the main army, but when seeing them run down the hill, the main force believed they were retreating and began embarking the ships. Under this disorder, the Scots charged at the beachhead and a terrible fight erupted. After hours of chaotic fighting, the Norwegians held out and managed to turn the tide. They drove the Scots beyond the hill and retook it, sending the Scottish army into withdrawal. It was a tactical draw, but a strategic loss. With the weather still volatile, King Haakon decided to retreat for the winter and continue with his campaign the following year.

The army returned to Kirkwall and wintered there. Interestingly, some Irish nobles met Haakon and asked for his military assistance in driving the English away from Ireland, proposing to take Haakon as *High King of Ireland* in return. Despite the enticing offer, Haakon rejected it. He would not betray his friend King Henry III of England, nor would he engage too heavily in foreign affairs.

Later that year, Haakon fell seriously ill. He lay in his bed for days without moving. He had Latin books read aloud for him, followed by all the Norse sagas. First the sagas of Norway's holy men and women, then *Fagerskinna*[33]. Haakon uttered few words. His health worsened by each page. Lastly, he had the saga of his grandfather Sverre read to him. It was obvious that Haakon would not live to see Christmas. On the 16th of December, he drew his last breath, aged 59.

---

[33] Another important collection of sagas about Norway's early kings and heroes.

Legacy

"Kingship was established and appointed to look after the needs of the whole realm and people... [The King's] chief business is to maintain an intelligent government and to seek good solutions for all the difficult problems and demands which come before him... But when there is no official business brought before him, he should meditate on the source of Holy Wisdom and study with attentive care all its ways and paths."

– extract from *King's Mirror.*

Haakon Haakonsson's body was brought back to Bergen, where his son and heir, Magnus, awaited him. It was a sad and emotional day when Haakon was buried in the Old Cathedral of Bergen. Unfortunately, in the wake of the Protestant Reformation, the Danes dismantled the Cathedral and destroyed its graves, including Haakon's.

Haakon Haakonson is as important as it is adventurous. He saved Norway from disintegration by restoring peace, stability, and growth. He found a Kingdom torn by chaos and civil war and left it in splendour and wealth. Haakon was clever and farsighted, setting in motion processes that Norwegians would benefit from years after his death. He was a visionary, defying the limits of the day and led Norway to greater heights than ever before. But Haakon was also a decision-maker who was not shy of taking responsibility or executing cunning plans. He had a deep and educated understanding of negotiation, psychology, and leadership, and would combine them in intricate, long-term strategies that slowly awarded Norway greater prosperity.

The sheer adventurous course of his life has also captured the imagination of many authors and artists: a child rescued from death by two, ski-wearing Birchlegs; a young prince hailed and celebrated by stout war veterans; a King who ruled with fervent energy and passion, building a nation out of a warzone. Quite the story. Henrik Ibsen, Norway's most famous playwright, wrote his thrilling play *Royal Pretenders* based on this story.

Haakon was not warlike, and his campaigns did not match up to the military performance of his grandfather Sverre. But this was arguably a blessing for Norway. Haakon was aggressive when needed, but never eager for war. He resolved most of his geopolitical challenges through calculated diplomacy, not war. He viewed war as a destructive force that should always be avoided, for it cost the lives of many intelligent and chivalrous men – lives that could never be "bought back". Therefore, Haakon often waited to see what mere words could achieve, and in his case, words were often sufficient.

His vast Norwegian realm was secured not through armies and invasions, but through cunning political manoeuvres. This saved many lives.

Haakon worked all his life to make his country a strong, civilised, morally adherent, and prosperous nation. He wanted to place Norway as an important and refreshing part of modern Europe. This vision did wonders: his reign was undoubtedly Norway's historical golden age.

Magnus the Law-Mender

1238 – 1280

*"From this you will observe that God demands moderation and fairness,
humility, justice, and fidelity as a duty from those whom he raises to honour,"*

– Haakon Haakonsson's advice to Norway's future Kings,
extracted from *King's Mirror*.

Who would want to be the successor to Haakon Haakonsson – a man with such an
astonishing track record? Who could live up to those expectations? Magnus *Lagabøte*
(the Law-Mender) Haakonsson knew how. He did not try to emulate his father's every
move. Instead, he would use his own talents and attributes to continue his father's
work by different means. Magnus' wise and responsible leadership, coupled with his
intellect, humility, and vision, would guide Norway to become one of the most modern
Kingdoms in the medieval world.

A Patient Little Brother

In Spring 1238, Magnus Haakonsson was born. His mother was the Queen of Norway,
Margarethe Skulesdottir, and his father, the King of Norway. He was therefore born into
a privileged world of royals: fine robes and capes, golden rings, lavish feasts, and servants
at every corner. Despite these comforts, Magnus seemed to have the rare ability to stay
humble and focused. This pleased his father, who carefully taught him the virtues of
humility, strength, and integrity.

Magnus was a Prince of Norway, but not Crown Prince, meaning he would not
be King. He had an elder brother who was the official heir, called Haakon the Younger.
This was a different character: he was athletic, confident and impulsive. His favourite
past time was hunting, and would visit hunting lodges across Scandinavia to hunt with
falcons, dogs and friends.

Prince Magnus savoured into books to study history and law. He underwent
strict and disciplined education and excelled at ease. Already as a teenager, Magnus

held advanced discussions with nobles from his father's court, often presenting original and new-thinking ideas. Out of all the subjects however, he favoured theology and philosophy. It is safe to say Haakon Haakonsson was blessed with two very talented sons.

In 1240, King Haakon made his Crown Prince, Haakon the Younger, co-ruler of Norway. This was to prepare the young man in administration, leadership, and kingship. Over the next fifteen years, he took young Haakon on campaigns and into heated negotiation chambers. He also sent him to be married to Rikissa Birgersdottir of Sweden, forging an important family union. But young Haakon would not enjoy marital bliss for long: while hunting in Sweden, he fell ill and, at the age of 25, died. It devastated Haakon Haakonsson and left a mourning Magnus, brotherless. During these sombre days, Magnus Haakonsson was announced the new heir to the throne.

From Crown Prince to King

Magnus was immediately put into leadership positions and drilled in kingship. He became the governor of Stavanger and Bergen, then later co-King of Norway – the second-in-command. Every time King Haakon was absent or away, Magnus had to be the effective ruler. He also joined his father in many negotiation sessions or banquets with foreign leaders, carefully observing how his father tactfully conducted himself. They probably had many, long conversations together, where Haakon taught him all that he knew, and Magnus would ask with deep curiosity and hunger to learn. The following words from *King's Mirror* are undoubtedly derived from Haakon's love to his son: *"It pleases me to hear you speak in this wise, and I shall be glad to answer; for it is a great comfort to me that I shall leave much wealth for my own true son to enjoy after my days."*

In 1261, Magnus married Ingeborg Ericsdottir, a Princess of Demark. The marriage was engineered by his father with the purpose of giving Norway a legitimate claim to the Danish throne. However, as Denmark was wrecked by civil wars, this marriage was never fully consented to by all Danish nobles, and the wedding ceremony had to take place in Bergen.

In 1262, Magnus bid goodbye to his father as he set sail from Bergen to Scotland. It would be the last time he saw his father. Some months later, Haakon's casket came sailing in the port. The time had come for Magnus to take the sacred title as King of Norway. The destiny of the nation depended on him. After his coronation, Magnus turned to his audience and swore by *lignum domini*[34] that he would give all of Norway one comprehensive justice system.

---

[34] Lignum domini was a splinter off the True Cross (see the chapter *Sigurd the Crusader*)

Consolidation

Scandinavia was changing. Denmark finally began stabilising from its anarchic civil wars and Sweden's military strength grew. However, Norway still seemed to be the great power of the North, as its borders stretched far and wide, with a rigid economy. King Magnus now had the responsibility to ensure the continuation of Norway's success story.

Magnus did not dare to gamble with his father's achievements in pursuit of grandiose ambitions. He lacked military verve, and therefore wanted to resolve his issues by other means. Magnus had other strengths, and he decided to adjust Norway's policies to them.

The war with Scotland came to an abrupt end in 1266, when King Magnus negotiated and concluded a peace treaty with King Alexander III of Scotland. The treaty ceded the Isle of Man and the Hebrides to Scotland – thus ending the territorial dispute. Thus, these islands, having been a part of the Norwegian world for hundreds of years, were formally absorbed into Scotland. However, this came with a price. The Scots agreed to pay 4000 marks sterling for four years *and* an "eternal" annual liability of 100 mark. In addition, they had to formally recognise Norway's legitimacy to rule over the Orkneys and the Shetlands. In other words, though the deal surrendered some islands, they formally secured many others *and* created a new source of income. Norway's possessions in the North Sea were consolidated.

Magnus' loyalty to peace was put to the test when the Danish King died and Magnus' strategically-given wife, Ingeborg, pushed her legitimate claim to the throne. The new Danish King counterargued her claims. Again, Scandinavia held its breath. However, Magnus desired peace. He abandoned the case and dropped the claims.

The New Justice System

"By Law shall the city be built."

– the preface of Magnus' Municipal Law code.

One of Magnus' chief goals was to make Norway much more advanced and efficient. The main instrument for doing so was the legislative. Personally, he had studied law all his life and saw the value of a logical, just, and sovereign judiciary. When his father produced *Haakon's New Law,* Magnus was one of the main clerks and advisors. *Haakon's New Law* was a great piece, but in Magnus' eyes it was unfinished. It had room for so much more.

Couriers were sent across the country to gather the most talented lawyers and diplomats. They all met in Bergen, in the royal estate, where they were embraced by the King himself. The most prominent members of Magnus' legal team were the baron Audun Hugleikson, an enormously wealthy baron with deep international experience; Lodin Lepp, a formidable diplomat who had established friendly relations with the rulers of Tunis and Egypt; Tore Bishop-son, an expert on canonical laws; Bjarne Lodinsson, a professor in Juris civilis at the University of Bologne with supplementary academic achievements from Paris. Their mission was clear: to develop a masterpiece within law and justice.

At this time, the legal system was based on several regional law-codes. This was the norm in the medieval world, but Magnus disagreed with it. He had meditated on his father's counsel in *King's Mirror*, where it warned: *"...for whenever the people are divided into many factions through loyalty to different chiefs, and these fall out, the masses will rashly pursue their desires, and the morals of the nation go to ruin. For then everyone makes his own moral code according to his own way of thinking; and no one fears punishment any longer when the rulers fall out and are weakened thereby."* This would inevitably lead to war and destruction, the *King's Mirror* argued. Having just risen out of 110 years of civil war, Magnus took this message to heart. He and his team of lawyers envisioned a nation where there was only *one, unified law-code* encompassing the *entire* realm and applying to *all* citizens. It would make the practice of justice much easier, unify the country further and produce a more stable, harmonious, and civilised society.

For such a national lawbook to be efficiently adopted and adhered to by the population, it had to be an organically produced – meaning, it had to have a firm foundation in current Norwegian customs and laws in all the things around the realm. Magnus and his team of legal experts went to work. They spent months collecting all the laws and noting all the customs in all the things, then brought them back to Bergen. There, they carefully studied and dissected the material, dismissing outdated or incoherent laws, and keeping the relevant ones. Having done this, they then went ahead to construct their own, new laws that they deemed necessary for Norway's advancement into modernity. Finally, in 1274, Magnus had his new law-code ready.

Before he could enforce it, he needed the formal ratification of the things. Remember, the tradition of things as the only institutions with legislative power was still largely intact. A King could compose laws and lawbooks, but he could *only* enforce them with the consent of the farmers and members of the things. So, Magnus contacted the largest things, the Gulathing, Frostathing, and so on. Only after receiving their consent could he finally enforce his new lawbook.

✴ Magnus the Law-Mender ✴

The new law-code was called *"Landsloven"* – the National Law-Code. It must have come as a surprise to the public, the aristocracy, and the clergy, as it was a revolutionary concept. One law-code to apply nationwide; one source of justice for all citizens alike – it was unprecedented. In fact, it was not only shocking in the North, but in all of Europe. By implementing *Landsloven*, Magnus made Norway world leading within the development and practice of modern law. Apart from the small kingdoms of Castille and Sicily, Norway became the *only* Kingdom in Europe that had one unified, national law. To illustrate how hyper-modern this idea was, lets keep in mind that the implementation of a national law-code did not arrive in Denmark until 1604, or France until *Code Napoleon* in 1800. Already in 1274, Norway had achieved this milestone under King Magnus.

Two years later, Magnus complemented *Landsloven* by publishing *Byloven* – the Municipal Law-Code. This code was introduced to promote urban planning and regulate the local conditions of Norway's major cities. It also outlined how houses ought to be built; streets to be organised; where marketplaces, docks, sewers and churches should be located; clarified how to buy and sell goods in the markets; what to do to prevent fire; and a range of other details to ensure that towns and cities were developed in an organised way. It made them easier to administer, and made them more responsive to local issues, bringing benefits specially to trade.

There were more revolutionary aspects to the National Code than just its nature. It itself provided new laws that changed Norwegian social norms. It commented on the rights of different social groups, marriage-laws, inheritance laws, taxation, and punishment, and further clarified the rules of succession. It emphasised King Haakon's ideology that the King and his court was to be the central *source* of all justice. On one hand, this finally abolished the right of private vengeance; but it also implied that, since the King was the main source of the law, he was *above* the law. For Norwegians, this was a major change. Norse tradition held that only the farmers and landowners of the thing-assemblies had legislative power, and that the King ought to be *under* the law. Interestingly, the English had followed this Norse tradition in 1215, when they placed their King *under* the Law via the *Magna Carta*. Ironically, in Norway, where this tradition had already held for centuries, Magnus the Law-Mender followed European feudalism by setting the King over the Law.

Both the National Code and the Municipal Code were written in a very systematic and well-formulated way, which reminds modern historians of how the Romans wrote their laws (it is very plausible that Magnus studied Roman history and law). This was yet another significant step forward, as thing-laws used to be ambiguous and often based on convention.

Political battles

King Magnus' new legal framework gave him much praise, but also enemies. A serious opponent of the King's reforms were the merchants of the Hanseatic league. The Hanseatic league was a commercial confederation of merchant guilds and towns in Northern Germany. Through aggressive trading strategies, these shrewd Germans gained increasing leverage over Northern Europe, and were now expanding into Scandinavia. However, King Magnus' municipal law-code stated that all who were commercially involved in Norway's cities ought to contribute to the city's expenses and challenges, regardless if that man was a foreigner or native. Not only did this add additional costs on Hanseatic accounts, but it also made them partly liable to Norwegian regulations – limiting their political agendas. It essentially gave King Magnus a leverage against them. Hansa merchants protested, but by doing so, revealed their inner intensions. This began to look more like a power-struggle than a trading relationship. Grievances deepened.

Another protester of Magnus' actions was the Archbishop of Norway, Jon the Red (Jon Raude). While Magnus composed the National Code, he had also revised the Christian Law of St. Olaf. This outraged the Archbishop, who insisted that only the clergy could address matters concerning Christendom and the Church, and that no other person, not even the King, had legal right to do so. By revising the Christian Law, Magnus implied he was *above the Church* – above God's representatives on Earth.

It is a paradox that Magnus clashed with the Church, because he was personally very devoted to Christian ethics. For example, throughout his reign, he worked extensively to improve the conditions of the disenfranchised and marginalised in society. His laws, for instance, made efforts to protect beggars, pregnant women, and orphans. He eagerly funded the Dominicans and Franciscans – two Catholic monk orders devoted to the hospitalisation, treatment, and aid of the poorest. Magnus partook in many of their projects too: rebuilding the damaged Cathedral in Bergen, holding several events to help the sick and needy, founding the Royal Chapel Organisation to organise the construction of chapels nationwide, and building low-priced hospitals for the poor. He remained a close ally and friend of the Franciscans until his death, and when he died, he was buried in the Franciscan Monastery. In foreign policy, he avoided war and guarded the peace.

But it did not stop the Archbishop in revolting. He published his own revised version of the Christian Law in defiance of the King, and illegally implemented it in Trondelag. Upon hearing this, King Magnus was exasperated. He demanded the Archbishop to comply to the King, but Jon flatly refused. The quarrel intensified.

After a while, Magnus realised the danger of this clash. It could trigger instability in the population and revive old feuds. Fearing this, Magnus chose to engage in a compromise. He brokered a concordat that granted the Church immunities (like immunity from conscription) and greater autonomy over their affairs. They were given a unique, independent Court to process clerical offenses and transgressors of the Christian Law (for example, clerical lawyers could prosecute and condemn adulterers, divorcers, and others). In other words, Magnus agreed to create another body of justice that focused solely on religious matters. In return, the Church agreed to not interfere in politics or succession procedures. Finally, a clear and defined separation of powers between Church and King was formed to avoid further disputes. The quarrel settled – but the grudge persisted.

"The Sharp Lamp of the Country"

Magnus continued with Haakon Haakonsson's quest to make Norway more European. He introduced new titles like "knight" and "baron", helping the prestige of the aristocracy take root. All men were to be greeted as "Sirs". He was the first to write his name with Latin numbers, stating he was Magnus the IV. In addition, he re-organised the Royal Guard, or hird (huscarls). His reign was characterised by peace and exponential growth. It was the pinnacle of Norwegian superiority. Bergen, its capital, was the trading metropolis of the North – gathering eager merchants from Greenland to Siberia; Iceland to Morocco. The merciless violence of the civil wars faded as old memories. The warmth of security, prosperity and order embraced the people. Norwegians lived good lives under this wise King. Magnus is personally described as an civilised, polite and amiable man. He cared more for justice than for power. He had the ability to listen carefully to specialists, advisors and others, and was very nuanced in his perspectives. He was beloved by his people.

After so many good and progressive years, Magnus "the Law-Mender" Haakonsson died in Bergen, 1280. He was buried in the Franciscan monastery, where he still lies today (this monastery is today known as *Bergen Domkirke* – Bergen Cathedral). Magnus' son, Eric Magnusson, was crowned as King – but it became soon clear that Eric was not like his father. Eric was known as Eric the Priest-Hater. In the saga of Bishop Arne it is written that: *"the rulership turned lopsided (...) when the sharp lamp of the country that King Magnus was, extinguished."*

Legacy

Magnus truly was a lamp of the country. He devoted his life and energy into the perfection of justice, and was not afraid to be controversial in doing so. He was highly educated, a pioneer within the legal system, and a zealous philanthropist. It is no denying that Magnus clearly loved his people and cherished his inheritance. Through his steadfast leadership, Norway became world-leading in the development of law and justice in Europe. The Kingdom was lifted to a higher plateau, behaving as the superpower in the North.

The Fall and Rise of the Kingdom

Haakon Haakonsson and Magnus the Law-Mender left behind a glorious inheritance. This put high expectations on their successors. Could they preserve the strength of the Kingdom? Who would become the "new Haakon" or "new Magnus"? The fate of the nation depended on it.

King Eric the "Priest Hater" Magnusson's reign was characterised by conflict, controversy, and ambition. During his 19-year-long reign, the Pope officially "gave up" Norway as a Catholic country and tensions resurged with Denmark. In 1299, Eric died without a son, passing the throne seamlessly to his brother, Haakon Magnusson.

Perhaps Norway's last pinch of glory and regional superiority came under King Haakon V Magnusson. He successfully produced a more proficient Norwegian military, made Oslo the permanent capital, built the impressive Akershus and Borhus Fortresses, conducted constructive reforms to consolidate the monarchy, further advanced and systematised Norwegian society and exerted significant political influence over Sweden. His daughter married a Swedish count and when she gave birth to a son, Magnus Ericsson, Haakon V placed little Magnus as his official heir. In July 1319, this Magnus Ericsson became King of Sweden, and, only a month later, Haakon V died, and Magnus inherited the throne of Norway as well. Haakon V had successfully placed his grandchild as King of both Norway *and* Sweden.

But although Magnus Ericsson was powerful, he lacked the decisiveness of his predecessors. After placing heavy tax burdens on the Swedes, a rebellion broke out where a German noble, Albert II of Mecklenburg, drove him from Sweden. Magnus tried to retaliate, but was defeated and he himself taken prisoner. His son Haakon VI the Younger succeeded as monarch in Norway.

Haakon VI the Younger took over as King of Norway in a period of sharp decline. Norway was weak. First and foremost, its economy had practically been hijacked by the Hanseatic League. These cunning tradesmen had cleverly gained monopoly in the

Norwegian corn and fish markets – in other words, the nerves of the Kingdom. By the 1300's, they even had their own mercenary army. The politico-economic influence they exerted over Norway was therefore unescapable. This was not unique to Norway – all of Scandinavia struggled to maintain integrity in the face of Hanseatic interests.

In the middle of this, Norway also experienced the relentless force of nature. In the late 1300s, the fruitful "medieval warm period" ended and was followed by the so-called "*little ice age*" – a period of much harsher and colder climates than before. Norwegian settlements on Greenland died out at this time. Due to Norway's vulnerable exposure to the North Sea and its very northern geographical position, the little ice age arguably rammed Norway harder than other Northern countries. Adding to the list of challenges was the gruesome Black Death. The horrific plague ravaged Norway very seriously. In just a few years, an estimated 60 percent of the entire Norwegian population died out (some say *two thirds*), completely sapping the Kingdom of manpower.

The panic spread among Norwegian nobles. Alf Erlingsson, the noble of Tonsberg, demonstrated Norway's futile desire to free itself from its decline. After years serving the King, Alf broke out of service, invaded the Hanseatic League but, lacking organisational skills and supplies, lost the campaign. Alf turned to piracy, trying to escape the heart-breaking sickness of Norway and building a success on his own. He was killed. The fall of the nobleman Alf Erlingsson is a symbol of the general direction Norway was heading.

While such tragic stories transpired, the top priority of King Haakon the Younger was not to solve Norway's critical problems right away. Instead, he spent his time trying to free his father from captivity. He forged an alliance with Valdemar IV of Denmark by marrying Valdemar's daughter, Margarethe, and mobilized the remnants of the Norwegian military. After a futile war, Haakon the Younger settled with a peace treaty. It was a disaster. Magnus Ericsson was released from captivity, but at an extraordinary price. The huge ransom was financed by a massive loan credited by the Hanseatic League, severely indebting the Kingdom of Norway and securing the Hanseatic grip on the country. The Hansa merchants secured another deal a few years later that granted them full monopoly over the most lucrative markets in Norway and Denmark. King Haakon the Younger had practically sold Norway to free his father.

Three years later, Magnus Ericsson died in an accident at sea. In 1380, Haakon the Younger died, leaving the throne to his young son with Margarethe, Olaf. When Olaf unexpectedly died in 1387, Margarethe became the undisputed Queen of Norway and Denmark.

Queen Margarethe proved to be incredibly efficient and authoritative, winning the love and respect of her subjects. When Swedish nobles asked her for help in evicting the

Germans from Stockholm, she invaded Sweden and won the Swedish throne as well. Thus, Queen Margarethe became the monarch of Norway, Denmark and Sweden. She was the absolute ruler of Scandinavia.

The idea of a united Scandinavia now began permeating the region. To bind all these kingdoms closer together, Margarethe gathered the foremost aristocrats of the North to the city of Kalmar. It was here, in 1397, that the famous Kalmar Union was forged – a royal union between the three Scandinavian countries. Margarethe's grand-nephew, Eric of Pomerania, was declared the first King of the Kalmar Union. Norway therefore officially lost its royal independency. It would not regain it until 1814.

Thanks to the Black Death and Little Ice Age, the Hanseatic League and a poor decision-making, Norway was in a serious crisis. It was much weaker than the other Scandinavian countries and failed to sustain the aggressive, confident policies it had enjoyed under Haakon Haakonsson, Magnus the Law-Mender and Haakon V. The military consisted of few men and outdated equipment. The economy was riddled with Hanseatic interference and debt. Sicknesses spread through the cities. Poverty increased. Norway gradually lost its strong voice in the courtrooms of the Kalmar Union.

In 1523 the Swedes broke out of the Kalmar Union after a series of disputes with Denmark. This was the end of the Scandinavian project and the beginning of the Denmark-Norway union. It was the final blow for Norway – the last nail in the coffin. Copenhagen was included into Central-Europe and enjoyed the many benefits this provided, while, by contrast, Oslo remained isolated in the shadow, cut off from European progress.

Exploiting the age of stagnation in Norway, the Danes gradually took steps of encroachment to firmly establish their overlordship. In the 1500's, the Danish King Christian III reduced Norway to a Danish province. A new social hierarchy emerged where the Norwegians always were at the bottom and the Danes at the top. Norwegians were rarely promoted to any position of authority, being rather used as simple workers. Few investments were made in Norway: schools and universities were not built; roads disintegrated; towns were riddled with dirt. All the while, the main recruitment grounds were in Norway – the Norwegians made up the core of the Danish army and navy. If a Norwegian sought any education, he had to travel to Copenhagen and attend schooling there. He had to learn Danish and always seek Danish approval for all he did. The Norwegians were locked in a dependency on Denmark. The Danes also took steps to quench Norwegian patriotism. The sagas were put away. Stories of the Norwegian titans faded. Monuments and landmarks dismantled and removed – including the grave of Haakon Haakonsson. Norwegians were encouraged to forget about St. Olaf, their eternal King.

Such were the conditions in Norway for four hundred years, until in 1814, when the handful of Norwegian soldiers, landowners and scholars gathered and declared Norway's independency from Denmark. They wrote a constitution and formed a sovereign government, all while reminiscing the stories of Norway's iconic foundation. This constitution preserved through to 1905, when Norway officially became fully independent. To honour the titans before them, the Norwegians voted in overwhelming majority to restore the ancient monarchy. Stories of their medieval heroes re-appeared. The Sainthood of Olaf the Stout re-emerged. The memory of the saga-Kings of Norway was passionately revived through Snorre Sturlason's writings. Emotions must have run deep for the Norwegian people when they experienced the crowning ceremony of King Haakon VII in 1906. The Kingdom of Norway was reborn.

> *"Raised is then once more within the boundaries of Norway the ancient throne which was occupied by Haakon the Good and Sverre, from which they ruled old Norway with wisdom and strength. That the wisdom and power exercised by them, the great kings of our ancient past, also will inspire the Prince which we, the freemen of Norway, in accordance with the wish of all the people, in gratitude and appreciation today unanimously have chosen, is a wish that every true son of Norway surely shares with me. God save old Norway!"*

– Georg Sverdrup, President of the Norwegian Assembly
for Independence in 1814.

Bibliography

Introduction

Den Svarte Vikingen; Birgisson, Bergsveinn; Spartacus, 2013.

The Penguin Historical Atlas of the Vikings; Haywood, John; Penguin Books, 1995.

Ingen Grenser; Heyerdahl, Thor; Lilliestrom, Per; JM Stenersens Forlag, 1999.

The Northmen's Fury - a History of the Viking World; Parker, Philips; Jonathan Cape, 2014.

Ynglingesaga; Norges Kongesagaer; Sturlason, Snorre; Forlaget LibriArte, 1997.

Norsk historie 800 - 1536; Sigurdsson, Jon Vidar og Anne Irene Riisøy; Samlaget, 2011.

The Conversion of Scandinavia; Winroth, Anders; Yale University Press, 2014.

Germanerne - Mytene, historien, språket; Janson, Tore; Pax Forlag, 2014.

The Sagas of the Icelanders; Kellogg, Robert; Penguin Classics, 2001

Den Svarte Vikingen; Birgisson, Bergsveinn; Spartacus, 2013.

The Penguin Historical Atlas of the Vikings; Haywood, John; Penguin Books, 1995.

Vikingenes tapte kongeriker; Griffiths, Jack; Aftenposten Historie. Nr. 4, 2016.

Harald Hårfagre; Titlestad, Torgrim; Saga Bok, Spartacus, 2012.

Outbreak of the Viking Age; Titlestad, Torgrim; Saga Forlag, 2018.

Harald Fairhair:

Harald Hårfagres saga; Norges Kongesagaer; Sturlason, Snorre; ForlagETT LibriArte AS 1997.

Vikinger og innbyrdeskriger i Norge; red. Neegaard, Dan Petter; red. Solberg, Finn Jørgen; Vega Publishing 2015.

Fagerskinna; red. Titlestad, Torgrim; Saga Bok 2008; 2.utg

✴ Bibliography ✴

Flatøybok bind 1; red. Titlestad, Torgrim; med.red. Rowe, Elizabeth Ashman; med.red. Birgisson, Bergsveinn; Saga Bok 2014

Harald Hårfagre; Titlestad, Torgrim; Saga Bok; Spartacus 2010.

Haakon the Good:

Håkon den godes saga; Norges Kongesagaer; Sturlason, Snorre; ForlagETT LibriArte AS 1997.

Vikinger og Innbyrdesstrider i Norge; red. Neegaard, Dan Petter; red. Solberg, Finn Jørgen; Vega Forlag 2015.

Fagerskinna; red. Titlestad, Torgrim; Saga Bok 2008; 2.utg

Håkon den Gode; Sigurdsson, Jon Vidar; Hellerud, Synnøve Veinan; Saga Bok; Spartacus 2012.

Flatøybok bind 1; red. Titlestad, Torgrim; med.red. Rowe, Elizabeth Ashman; med.red. Birgisson, Bergsveinn; Saga Bok 2014

Ingen Grenser; Heyerdahl, Thor; Lilliestrom, Per; JM Stenersens Forlag AS 1999.

Norsk historie 800 – 1536; Sigurdsson, Jon Vidar; Riisøy, Anne Irene; Samlaget 2011.

Olaf Tryggvason:

Olav Tryggvasons saga; Norges Kongesagaer; Sturlason, Snorre; ForlagETT LibriArte AS 1997.

Fagerskinna; red. Titlestad, Torgrim; Saga Bok 2008; 2.utg

Flatøybok bind 1; red. Titlestad, Torgrim; med.red. Rowe, Elizabeth Ashman; med.red. Birgisson, Bergsveinn; Saga Bok 2014

Flatøybok bind 2; red. Titlestad, Torgrim; med.red. Rowe, Elizabeth Ashman; med.red. Birgisson, Bergsveinn; Saga Bok 2015

Olav Tryggvason; Tjønn, Halvor; Saga Bok; Spartacus 2012.

Håkon jarl; Stylegar, Frans-Arne; Saga Bok; Spartacus 2013.

Millennium; Holland, Tom; Abacus 2009

Vikinger og Innbyrdesstrider i Norge; red. Neegaard, Dan Petter; red. Solberg, Finn Jørgen; Vega Forlag 2015.

The Northmen's Fury – a History of the Viking World; Parker, Philip; Jonathan Cape 2014.

✳ Bibliography ✳

Ingen Grenser; Heyerdahl, Thor; Lilliestrom, Per; JM Stenersens Forlag AS 1999.

The Penguin Historical Atlas of the Vikings; Haywood, John; Penguin Books Ltd.; 1995.

The Complete Sagas of the Icelanders; Leifur Eriksson Publishing Ltd; Penguin Books 2001.

Saint Olaf

Olav den helliges saga; Norges Kongesagaer; Sturlason, Snorre; ForlagETT LibriArte AS 1997.

The Northmen's Fury – A History of the Viking World; Parker, Philip

Fagerskinna; red. Titlestad, Torgrim; Saga Bok 2008; 2 utg.

Olav den hellige; Titlestad, Torgrim; Saga Bok; Spartacus 2013.

Flatøybok bind 2; red. Titlestad, Torgrim; med.red. Rowe, Elizabeth Ashman; med.red. Birgisson, Bergsveinn; Saga Bok 2015

Olav den hellige; Langslet, Lars Roar; Gyldendal Forlag 1995.

Tanker Om Tro; Langslet, Lars Roar; St. Olav Forlag 2013.

Saint Olav – King of Norway; Müller, Olav; Maximilian og St. Olav Forlag

I Saw It Happen In Norway; Hambro, Carl Joachim; D. Appleton-Century Company Inc. 1940.

Magnus the Good:

Magnus den godes saga; Norges Kongesagaer; Sturlason, Snorre; ForlagETT LibriArte AS 1997.

Vikinger og innbyrdeskriger i Norge; red. Neegaard, Dan Petter; red. Solberg, Finn Jørgen; Vega Publishing 2015.

Millennium; Holland, Tom; Abacus 2009

Magnus den gode; Morten, Øystein; Saga Bok; Spartacus 2011.

Fagerskinna; red. Titlestad, Torgrim; Saga Bok 2008; 2 utg.

Harald Hardrada:

Harald Hardrådes saga; Norges Kongesagaer; Sturlason, Snorre; ForlagETT LibriArte AS 1997.

✷ Bibliography ✷

Vikinger og innbyrdeskriger i Norge; red. Neegaard, Dan Petter; red. Solberg, Finn Jørgen; Vega Publishing 2015.

Millennium; Holland, Tom; Abacus 2009

The Battle of Hastings; Wood, Harriet Harvey; Atlantic Books; Grove Atlantic Ltd 2009.

The Penguin Historical Atlas of the Vikings; Haywood, John; Penguin Books Ltd.; 1995.

The Northmen's Fury - a History of the Viking World; Parker, Philip; Jonathan Cape 2014.

Fagerskinna; red. Titlestad, Torgrim; Saga Bok 2008; 2 utg.

Harald Hardråde; Tjønn, Halvor; Saga Bok; Spartacus 2010.

Olav Kyrre; Bandlien, Bjørn; Saga Bok; Spartacus 2011.

Olaf the Peaceful and Magnus Barefoot:

Magnus Barfots saga; Norges Kongesagaer; Sturlason, Snorre; ForlagETT LibriArte AS 1997.

Olav Kyrres saga; Norges Kongesagaer; Sturlason, Snorre; ForlagETT LibriArte AS 1997.

Vikinger og innbyrdeskriger i Norge; red. Neegaard, Dan Petter; red. Solberg, Finn Jørgen; Vega Publishing 2015.

Magnus Berrføtt; Føresund, Randi Helene; Saga Bok; Spartacus 2012.

Olav Kyrre; Bandlien, Bjørn; Saga Bok; Spartacus 2011.

Sigurd the Crusader:

Magnussønnenes saga; Norges Kongesagaer; Sturlason, Snorre; ForlagETT LibriArte AS 1997.

Jakten på Sigurd Jorsalfare; Morten, Øystein; Spartacus Forlag 2014

Fagerskinna; red. Titlestad, Torgrim; Saga Bok 2008; 2 utg.

Ingen Grenser; Heyerdahl, Thor; Lilliestrom, Per; JM Stenersens Forlag AS 1999.

Sigurd 1 Magnusson Jorsalfare; Krag, Claus; Norsk Biografisk Leksikon 2009.

Kong Sigurds Jorsalferd; Bergan, Halvor; Norgesforlaget 2005.

Øystein 1 Magnusson; Krag, Claus; Norsk Biografisk Leksikon 2009.

✳ Bibliography ✳

Sverre:

Sverres saga – en tale mot biskopene; I. M. Stenersens Forlag 1914.

Sverresoga; red. Halvdan Koht; Det Norske Samlaget 1995.

Norsk historie 800 – 1536; Sigurdsson, Jon Vidar; Riisøy, Anne Irene; Samlaget 2011.

Kong Sverre – Norges største middelalderkonge; Krag, Claus; H. Aschehoug & Co 2005.

Våre Konger; Langslet, Lars Roar; J.W. Cappelens Forlag 2002.

Sverre Sigurdsson; Helle, Knut; Norsk Biografisk Leksikon 2009.

Haakon Haakonsson the Great:

Håkon Håkonssons saga; Thordarson, Sturla; Det Norske Akademi for Sprog og Litteratur; Thorleif Dahls Kulturbibliotek; Aschehoug 2008.

Norsk historie 800 – 1536; Sigurdsson, Jon Vidar; Riisøy, Anne Irene; Samlaget 2011.

Håkon 4 Håkonsson; Helle, Knut; Norsk Biografisk Leksikon 2009.

Vikinger og Innbyrdesstrider i Norge; red. Neegaard, Dan Petter; red. Solberg, Finn Jørgen; Vega Forlag 2015.

Våre Konger; Langslet, Lars Roar; J.W. Cappelens Forlag 2002.

Magnus the Law-Mender:

Magnus 6. Håkonsson Lagabøte; Helle, Knut; Norsk Biografisk Leksikon 2009.

Norsk historie 800 – 1536; Sigurdsson, Jon Vidar; Riisøy, Anne Irene; Samlaget 2011.

Våre Konger; Langslet, Lars Roar; J.W. Cappelens Forlag 2002.

Håkon Håkonssons saga; Thordarson, Sturla; Det Norske Akademi for Sprog og Litteratur; Thorleif Dahls Kulturbibliotek; Aschehoug 2008.

Med lov skal byen bygges; Høiaas, Knut; Bryggens Museum.

King's Mirror; Haakonsson, Haakon; translated by Laurance M. Larsen, 1917 webarchive.org.

Printed in Great Britain
by Amazon

27223910R00099